PSYCHOLOGY LIBRARY
EDITIONS:
PERSONALITY

Volume 13

APPROACHES TO
PERSONALITY THEORY

APPROACHES TO PERSONALITY THEORY

DAVID PECK AND DAVID WHITLOW

Routledge
Taylor & Francis Group

LONDON AND NEW YORK

First published in 1975 by Methuen & Co. Ltd

This edition first published in 2019
by Routledge
2 Park Square, Milton Park, Abingdon, Oxon OX14 4RN

and by Routledge
52 Vanderbilt Avenue, New York, NY 10017

Routledge is an imprint of the Taylor & Francis Group, an informa business

British Library Cataloguing in Publication Data
A catalogue record for this book is available from the British Library

ISBN: 978-0-367-03112-1 (Set)
ISBN: 978-0-429-05756-4 (Set) (ebk)
ISBN: 978-0-367-13587-4 (Volume 13) (hbk)
ISBN: 978-0-429-02734-5 (Volume 13) (ebk)

Publisher's Note
The publisher has gone to great lengths to ensure the quality of this reprint but points out that some imperfections in the original copies may be apparent.

Disclaimer
The publisher has made every effort to trace copyright holders and would welcome correspondence from those they have been unable to trace.

APPROACHES TO PERSONALITY THEORY

**David Peck and
David Whitlow**

Methuen

First published 1975 by Methuen & Co Ltd
11 New Fetter Lane, London EC4P 4EE
© 1975 David Peck and David Whitlow
Printed in Great Britain by
Richard Clay (The Chaucer Press), Ltd
Bungay, Suffolk

ISBN (hardback) 0 416 82800 0
ISBN (paperback) 0 416 82810 8

We are grateful to Grant McIntyre of
Open Books Publishing Ltd for assistance
in the preparation of this Series.

Contents

Editor's Introduction

David Peck and David Whitlow provide a thorough and fair-minded review of the major personality theories in psychology. A noteworthy feature of their book is their review of the techniques of assessment which are associated with each theory; also of great value is their selection of areas of application to practical problems. One is left with the feeling of amazement at the thought and effort which have been expended by psychologists and yet the extent of our ignorance.

Unit D is a crucial part of *Essential Psychology*. Many who are dissatisfied with all current models of man see the concept of man as an individual and social person as the best alternative. This is because it emphasizes the uniqueness of the experience of each individual and also the notion that he acts upon his environment in a purposeful way. The books in this unit all demonstrate how the basic assumptions of personality theory and research are changing. Instead of personality being described in terms of forces driving people from within or events manipulating them from without, individuals are now being described as persons each with his own way of construing reality.

Essential Psychology as a whole is designed to reflect the changing structure and function of psychology. The authors are both academics and professionals, and their aim has been

7

to introduce the most important concepts in their areas to beginning students. They have tried to do so clearly but have not attempted to conceal the fact that concepts that now appear central to their work may soon be peripheral. In other words, they have presented psychology as a developing set of views of man, not as a body of received truth. Readers are not intended to study the whole series in order to 'master the basics'. Rather, since different people may wish to use different theoretical frameworks for their own purposes, the series has been designed so that each title stands on its own. But it is possible that if the reader has read no psychology before, he will enjoy individual books more if he has read the introductions (A1, B1, etc.) to the units to which they belong. Readers of the units concerned with applications of psychology (E, F) may benefit from reading all the introductions.

A word about references in the text to the work of other writers – e.g. 'Smith, 1974'. These occur where the author feels he must acknowledge by name an important concept or some crucial evidence. The book or article referred to will be listed in the References (which double as Name index) at the back of the book. The reader is invited to consult these sources if he wishes to explore topics further. A list of general further reading is also to be found at the back of this book.

We hope you enjoy psychology.

Peter Herriot

I
Introduction to personality theory

In everyday language we often refer to another person by saying 'He has no personality', or that 'she has a strong personality'. We would understand this to mean that he is boring, unimaginative and predictable, and that she is domineering, rigid, reluctant to change her mind and so on. Psychologists, however, use the term 'personality' in a much more restricted and technical sense. In a psychological sense, a strong person has no more personality than a weak one, in the same way that a bright red coat has no more colour than a drab brown one. All things have colour; all people have a personality. They differ not in the amount they have, but in the type they have. The general use of the term personality is, therefore, very different from that of psychologists.

What then do psychologists mean by the term personality? Here we immediately get on to difficult ground. There is no single, generally accepted use of the term amongst psychologists; indeed, some would maintain that in the sense that it is ordinarily used by psychologists there is no such thing as personality (as we shall see in later chapters). Psychologists have adopted a bewildering variety of approaches to, and an inevitably equally bewildering array of definitions of, the concept of personality. Before we go on to examine the reasons for this, let us first form at least a rough idea of what psychologists mean by the term.

Most theorists accept that the major, if not the only, way to study personality is by observing what people actually do or say that they do; either directly by watching someone behave, and tallying what he does, or indirectly by counting up his responses to questionnaire items. A basic assumption is that the term personality relates to what people do or what they experience. A second frequent assumption is that personality is an entity; that is, it really exists and is not just a convenient short-hand way of summarizing a person's behaviour. A third assumption is that personality is relatively fixed and enduring, so that the 'core' remains relatively immutable, while only its more surface features are modifiable. Due acknowledgement is sometimes, but not always, paid to the moderating influence of the situation in which the behaviour takes place.

Armed with these assumptions, most personality theorists have been content with such definitions of personality as 'those structural and dynamic properties of an individual as they reflect themselves in characteristic responses to situations' (Pervin, 1970); or 'those relatively stable and enduring aspects of the individual which distinguish him from other people and, at the same time, form the basis of our predictions concerning his future behaviour' (Wright *et al.*, 1970); or 'a dynamic organisation within the individual of those psychological systems that determine his characteristic behaviour and thought' (Allport, 1961). Some psychologists have defined 'personality' very widely so that it covers virtually everything and anything that a person does, from how he solves problems and how he deals with incompatible thoughts, to changes in physiological functioning in response to emotion-arousing situations. Although it may be reasonable to adopt such positions, most psychologists would concur with the type of definitions given above, emphasizing the individual's patterns of behaviour within a social or interpersonal context. Personality theorists have maintained that for any behaviour to be of interest it must be characteristic of the person (i.e.

10

differentiate him from other people) and it should manifest itself in a variety of situations and be consistent over time.

Given that personality theorists have the same subject matter (man), are interested in the same aspects (man's behaviour), and are largely interested in the same goals (description, understanding and prediction), let us now consider why psychologists have produced so many different definitions and approaches. The main reason must surely be that human behaviour is enormously complex, and is determined by not one factor or even one set of factors, but by a vast number of determinants at many different levels. With such a complex sphere of investigation, each different approach has emphasized different kinds of observations about people, which are inevitably reflected in different kinds of personality theories.

This complexity of subject is revealed by a glance at the different areas of personality which have been detailed (sometimes quite arbitrarily) by psychologists, such as personality development, personality dynamics, personality assessment, personality structure, personality change, personality adjustment, and so on. It is rather as if ten blind men surrounding an elephant were reaching out and touching the nearest part, then theorizing about what it was. Each individual's description and theorizing might be perfectly adequate, but would tell us little about the overall function and structure of the complete elephant.

Not only do personality theorists sample different aspects of human behaviour, but they also use different tools to do so. Some theorists have used samples of unstructured speech as a major investigatory tool; others have used sophisticated mathematical analyses of responses to paper-and-pencil questionnaires; some have simply counted items of behaviour, and still others have used the size and shape of the body. There is, therefore, no such thing as a theory of personality, in the sense that a theory covers all aspects of human behaviour, but there are many theories whose main area of interest lies within the domain of personality.

What determines the aspects examined and tools adopted by any particular theorist may result in part from his own

11

implicit ideas of what human behaviour is all about. What one believes man to be will determine which aspects of human behaviour are important and are to be studied, and which are unimportant and can be safely ignored. A related topic is, what are the functions of a personality theory, and it is to this that we shall now turn our attention.

Functions of personality theories

One function of scientific theories is to organize accumulated knowledge in a given area into a form which renders it usable and communicable to other people. Theories are useful ways of fitting facts together, permitting one to generate hypotheses and to apply the organized knowledge to new circumstances and problems as they arise. Theories should enable one to predict the future and explain the past and present by reference to the propositions contained within them. What is a useful theory at one period may not be so at a later period; a theory may have to be abandoned and replaced by a new one which better accounts for the observed facts (see F7 of *Essential Psychology*). However, to abandon a theory is not to deny the validity of the observations upon which the theory was based. The theoretical foundations of physics have undergone enormous re-thinking of late, such that the whole theoretical structure is being constantly undermined; however, when we turn a lightswitch, the light still comes on. Essentially, a theory is only a useful way of ordering and simplifying facts into general laws or propositions.

Personality theories serve the same general purposes as other types of scientific theories. The specific functions of a theory of personality are broadly agreed upon by different theorists. Most would maintain that the theory must be based on important phenomena, that it should parsimoniously account for the facts, should stimulate further research, should incorporate known empirical findings, should simplify the complexities of human behaviour and so on. Levy (1970) considers that all these criteria fall into three broad over-

12

lapping categories, the importance attached to each being largely a matter of personal judgement. His criteria were grouped as follows: subjective-judgemental, logical-epistemic and empirical.

Subjective-judgemental criteria
As the title suggests, such criteria may vary from one person to another; no objective standard can be applied, but this does not necessarily reduce the importance of the criteria.

Personality theories vary, for example, in the 'extensiveness of the domain' of the phenomena covered, and in the 'importance of the domain', and these can be used as bases upon which to judge a theory. There is, however, no final arbiter of extensiveness or importance. Thus, some theories of personality have been taken to task because they may not incorporate thinking, cognitive styles, drive states, and so on; whereas other theories have been criticized because they have offered few practical solutions to important human problems. Clearly these criteria are applied according to the personal preference of the judges, and the subjective basis of such judgements should be recognized. In addition, these criteria should be viewed in the context of what the personality theorist is trying to do. To criticize a theorist who is interested in the physical basis of personality on the grounds that his theory has little to say about speed of learning is perhaps unfair; whereas to lodge the same criticism against a theorist who is interested in speed as an aspect of personality would be more justifiable. Thus we must distinguish between criticisms of what a theory is claimed to be about, and criticisms of what it is believed a theory ought to be about.

Logical-epistemic criteria
Such criteria demand that a theory be stated in sufficiently explicit terms to enable testable predictions to be made. A well-constructed theory should enable logical deductions to be made, and empirical answers to be found. Accordingly, the theory should not be such that contrary hypotheses about the same phenomena are derivable from it, but should be logic-

13

ally consistent, with each proposition or law being fully compatible with every other.

If testable deductions can be made, it follows that in the event of incompatible findings, the theory must be capable of modification; that is, it must be responsive to data. It is the essence of theories that they should change in the light of new facts. The theory is embellished, enriched or abandoned according to these new facts and its usefulness depends in part upon its ability to incorporate new evidence.

A further criterion is the amount of research which is stimulated by a theory. Indeed, to quote Hall and Lindzey's (1957) influential textbook *Theories of Personality*, 'the fruitfulness of (personality) theories is to be judged primarily by how effectively they serve as a spur to research' (p. 27).

Empirical criteria

A theory may be extensive, cover important areas, be internally consistent, and lead to testable predictions; but it is argued here that the acid test is whether the theory can fulfil empirical criteria. If it cannot, then it is interesting but rather pointless. What do we mean by empirical criteria? Basically, we mean how well does the theory work, in two main respects – its validity and its utility. We assess *validity* when an attempt is made to support a concept or deduction from a theory by comparison with information obtained from another independent source such as an experiment. We may, for example, predict from a theory that there is a positive relationship between extraversion and susceptibility to the effects of alcohol. To test this, we could give a standard amount of alcohol to people with different degrees of extraversion, and assess the effects of alcohol on, for example, the ability to perform mental arithmetic or a hand-eye co-ordination task. If the theory is correct, intoxicated extraverts should produce fewer correct answers to the arithmetic problems, and should display poorer hand-eye co-ordination. When such investigations are supportive of a theory, it should not be assumed that the theory as a whole has been shown to be valid, but merely that the prediction has been upheld in a

14

certain area. Furthermore, if the prediction is not upheld, this does not mean that the whole theory is untenable, but that the one aspect of it may require some modification.

Utility refers to the value of the theory to the user. Personality theories might reasonably be expected to provide information useful in the treatment of mental disorders, in the selection of personnel, in understanding the behaviour of schoolchildren, or in the assessment of criminals; that is, they are expected to be useful in the resolution of practical problems. Utility is obviously partly a function of most, if not all, of the preceding criteria, but these alone do not guarantee the theory's utility. A theory may, for example, predict with acceptable degrees of accuracy the job most suitable for a person with a particular personality type. However, simply asking the person might have produced the same information much more quickly and at a fraction of the cost. In other words, the value or utility of a theory should be judged in relation to alternative ways of arriving at the same decision.

All of these criteria for the evaluation of a personality theory are important. However, if a theory is to be taken seriously, and not regarded as a purely academic exercise far removed from the exigencies of the real world, then its validity and utility are of paramount importance. Psychologists should not indulge in the rather incestuous practice of devising and refining personality theories for consumption by other psychologists, without consideration of how the theory can be applied. After all, personality is about human behaviour, which typically takes place outside psychological laboratories. The importance of applying theories derived in the laboratory to outside situations is neatly illustrated in the following experiments: A series of laboratory studies by Asch (1958) (see B1) demonstrated that about 30 per cent of his subjects conformed to, and agreed with, the clearly incorrect judgements of a number of 'stooges' in an experiment concerned with estimation of the comparative lengths of lines. Later studies by Luchins and Luchins (1967), however, were unable to produce the same amount of conformity when the experiment was repeated in a more 'natural setting',

15

such as an airport terminal. Thus, unless theories are tested in more realistic and natural settings, psychologists could well end up with a series of experiments providing us with knowledge of human behaviour in the laboratory, which relates little to human behaviour elsewhere (see F8).

A compelling reason for using utility as a major criterion in assessing personality theories is that most theorists make extensive claims for the practical utility of their theories; it is important to assess how far these claims are justified. It is, of course, possible that although a theory may be of little current utility, useful applications may well be found at a later date.

Naïve theories of personality

Why is it necessary for psychologists to devise personality theories? People are continually making assessments of other people, judging from personal appearance and mannerisms, listening to what others say and watching what they do in various contexts, and then relating such observations to their own implicit theories of personality, inferring enduring dispositions and underlying motivation. There is, however, a body of evidence indicating that subjective assessments of another's behaviour are extremely liable to error, and that even in a reasonably well-structured situation, such as an interview, the accuracy of assessments is quite minimal.

For example, our subtle expectations about another person may have a marked effect on our assessment of him and on our own behaviour towards him. Kelley (1950) demonstrated this in the following experiment: a male lecturer was introduced to a class of male students as 'Mr .... is a graduate student in the Department of Economics and Social Science. He had three semesters of teaching experience in psychology at another college. He is 26 years old, a veteran, and married. People who know him consider him to be a rather cold person, industrious, critical, practical and determined.' Half of the students received the above description, half of them a similar description, except that the words 'rather warm' were substituted for 'rather cold'. Afterwards, the students were required to rate the lecturer; more favourable ratings were made by the students who expected him to be 'warm'; in addition 56 per cent of the students given the 'warm' expectation joined in the discussion after the lecture, whereas only 32 per cent of the 'cold' expectation students joined in.

It has also been shown that our friendship choices are not normally based upon a fully impartial assessment of the other's qualities and behaviour. A very important determinant of whom we shall like seems to be how often we interact with that person. For example, Festinger, Schachter and Back (1950) examined friendship formations in a housing estate, and found, as one might expect, a high relationship between friendship choice and physical distance. But physical distance was not the only important factor; the precise situation of an apartment was also important, in that people on the ground floor were likely to have friends on the upper floor if their apartment was near the foot of the stairs, but not if it was nearer the centre of the row of apartments. This seemed to be because people in apartments near the stairs had many more opportunities to meet and make 'passive contacts'; they concluded 'the relationships between ecological and socio-metric (i.e. friendship choice) structures is so very marked that there can be little doubt that in these communities passive contacts are a major determinant of friendship and group formation'. Warr (1964) has replicated some of these findings in a British university men's hall of residence.

Not only do we sometimes use imprecise and possibly misleading criteria in judging others, but we are also prone to accept as useful and valid descriptions of behaviour which are so general that they could apply to anyone. In one study (Ulrich, Stachnik and Stainton, 1963) university students were given personality tests, followed by a written interpretation allegedly based on the test results. In fact, each interpretation was identical, containing such statements as 'you prefer a certain amount of change and variety and become dissatisfied when hemmed in by limitations and restrictions ... You pride

17

yourself as being an independent thinker and do not accept others' opinions without satisfactory proof ... Some of your aspirations tend to be pretty unrealistic'. Almost all the students reported that the interpretations were accurate descriptions of their behaviour, some of them claiming that their personal problems had been helped. Subsequent studies have shown a similar degree of acceptance of generalized personality descriptions, even when given by a non-creditable source, and when the descriptions were unfavourable. These experiments on the 'Barnum effect', as it is called, are not important as demonstrations of credulity, but are important in illuminating the dangers of vague personality descriptions and naïve interpretations. Clearly if psychological descriptions of personality are to be taken seriously, they must do better than this.

Psychologists have devised many personality theories; some theories have stood the test of time, others have had a short but explosive life, only to fade away, and some have been taken up and modified by other theorists so that the original forms are barely recognizable; many are of purely historical interest. It is impossible to survey all the personality theories put forward by psychologists, and some criteria has to be adopted for selection. It is intended that the theories discussed in this book should be representative of current psychological thinking in the area of personality; one objective way of selection for discussion is to note the impact of the theories as reflected in references to them in the recent psychological literature. The theories included here appear to be the most vigorous, stimulating and potentially fruitful approaches to the domain of personality. Others have been included, however, not necessarily because of their current status, but because of their highly influential stature in relation to later developments.

One main way in which personality theories can be classified is in terms of their comprehensivenes (Hall and Lindzey, 1957); that is, do the theories attempt to encompass and describe the whole of a person and his behaviour and attempt to predict behaviour in a wide range of settings; or do they attempt to cover only specific content areas, and relate to behaviour only in a narrow range of circumstances? Although it is possible to argue that the latter are not really theories of personality at all, since they cover such limited aspects of human behaviour, nevertheless these 'narrow-band' theories are of increasing importance and they are highly typical of the work that is currently being conducted in the area of personality, both in research and in applied settings. Accordingly, the narrow-band theories have been given due representation in Chapters 5 and 6.

Theory and assessment in personality must go hand-in-hand. Some form of measurement technique must be devised in relation to a particular theory; without this, the theory cannot be developed and assessed. The act of measurement can, in turn, help to promote the theory. Accordingly, the measurement techniques in relation to each personality theory will be examined in some detail, particularly in respect of the reliability of the techniques.

Reliability refers to how far a measuring device gives similar readings when it is assessing the same thing under slightly different circumstances. There is more than one kind of reliability, but the one of major concern here is reliability over time, or test-retest reliability. To assess test-retest reliability, one would administer a test to a large number of people on two occasions, separated by an interval of time; if the test is reliable, similar scores on the test should be obtained on each occasion. Another type of reliability is inter-observer reliability, which refers to the degree of concordance between two or more people when they are assessing the same events. The degree of reliability is normally expressed in terms of a 'correlation' (see A9). (A correlation is a statistical way of expressing the degree of relationship between two things; a high positive correlation would be in the range of ·7 to 1; a moderate correlation from ·4 to ·7; a low correlation from ·2 to ·4; and a negligible correlation from 0 to ·2.)

The concept of *validity* has already been briefly discussed. The validity of a personality test refers to how far the test actually measures what it is supposed to measure. There are many different kinds of validity, but in assessing the utility of a personality test, empirical validity is particularly important. Empirical validity is determined by examining the relationship between test scores and another criterion of the behaviour under investigation. Scores on a test of neuroticism, for example, should differ when administered to clinically neurotic subjects in a hospital, and when administered to subjects selected randomly from the general population. Validity is also often expressed in terms of a correlation.

Although personality theories cannot easily exist without related measurement techniques, the converse does not seem to be the case; there are many examples of personality measures which are not related to an explicit theory of personality. Goldberg (1971) has stated 'new personality scales and inventories are at least as likely to be focused upon constructs arising out of applied societal pressures as upon any theories of personality'. Although such personality assessment procedures are clearly of importance, measurement devices not related to our major concern of presenting and evaluating specific personality theories will not be covered.

2
Psychoanalytic personality theory

There is no single version of psychoanalytic personality theory that can be designated as the 'true' version; Freud frequently claimed that with conceptual, methodological and technological advances quite major changes in emphases and even radical transformation of crucial concepts would occur, and that such occurrences would be welcomed. Throughout his life, Freud described some quite major changes in his thinking in the light of accumulating clinical experience. Unfortunately, it was sometimes left unclear whether the earlier statements should henceforth be disregarded and considered to be superseded, or whether later versions should be regarded as being simply complementary to the earlier.

Although such versions are rarely if ever completely contradictory, some confusions have inevitably arisen, and there does not seem to be any authoritative integration of developments. Some writers still place a great deal of emphasis on Freud's division of the systems of the mind into the Unconscious, Preconscious and the Perceptual-Conscious, which others regard as being unimportant and largely superseded by the later developments. Further confusion was added by Freud's use of the same term to refer to different things at various different stages in his thinking. Because of this confusion, it is possible that some errors of omission or commission may be found in this necessarily brief presentation.

Personality structure

In Freud's theory, the structure of the mind can conveniently be seen as comprising three main parts, differentiated by their functions. These are the id, ego and superego.

The id. The fundamental source of personality is the id; it incorporates all the inherited characteristics of the individual, especially the instincts, and is closely related to biological functions such as breathing, sweating and elimination. Further, it is the reservoir of all the energies utilized by the mind and the body. A primary function of the id is to maintain a balance in the forces which pull people in various ways. These forces produce a tension which the id cannot tolerate, and it takes immediate steps to reduce the tension and restore the balance. The id functions in an irrational, impulsive way, paying little heed to possible social restrictions; this tendency towards immediate gratification is called the 'Pleasure Principle'. The main force of the instinctual drive of the id is the 'Libido', a force of energy whose nature is primarily sexual, but which also serves to motivate self-preservation. (It should be noted that, for Freud, 'sexual' meant a great deal more than overt sexual activity; all forms of pleasurable sensations, affection, friendship and so on were encompassed by this term.)

In contrast to the libidinal instinct, Freud also described the instinct of Thanatos (or a death instinct) to account for mankind's innate aggression and destructiveness, for which Freud found evidence in the slaughter of the First World War. The death instinct is originally directed towards the individual himself, and manifests itself by suicide, self-injury and sado-masochistic behaviour, or by externalizing the instinct onto other people. The instinct of Thanatos aroused a great deal of controversy and has never been fully accepted by many otherwise orthodox Freudians.

Tensions build up in the id and demand release. When

there is a need to reduce tension the id conjures up an internal fantasy of the object of the instinctual drive (e.g. the mother) and achieves vicarious satisfaction by releasing the tension in relation to the fantasy rather than to the real-life object. The id is not therefore capable of complete tension reduction since it operates only by means of fantasy, and has no direct access to the external world.

The ego. Since the id cannot achieve direct discharge of instinctually-derived tensions, this function is taken over by another major structure of the psyche – the ego. The ego is an offshoot of the id and grows directly from it, but operates independently. Its main function (called the Reality Principle) is to regulate the id's irrational frenetic drives for fulfilment, and to channel the drives into more acceptable forms of expression, continually seeking the best compromise between the constraints of the outside world and the drives of the inner world. The ego has to scan the real world to locate objects upon which the drives can be appropriately released, and is in close contact with reality. It acts as a kind of dam, restraining the pressure from the reservoir of the id until the environment provides acceptable conditions for release.

The ego achieves this by a variety of techniques, but chiefly by displacement. The ego could not, for example, allow the direct discharge of aggressive impulses towards one's father; accordingly, the cathexis (or the recipient object of the instinctual drives) may be displaced onto authority figures in general, producing instead criminals or 'drop-outs'.

There are occasions when the ego is caught off guard, and at such time the impulses from the id almost reach consciousness, but still in a disguised form. Such events can lead the trained observer to discover much about the inner strivings and impulses; thus, slips of the tongue, errors in writing, dropping and breaking objects and so on – these are all *determined* by the psyche and are not simply due to chance factors. Similarly, jokes, dreams and responses to ambiguous stimuli can provide clues to the unconscious strivings. The interpretation of dreams in particular is a major technique

23

involved in therapeutic methods based on psychoanalytic theory.

When the ego is not adequately controlling the impulses of the id, anxiety develops because the person may begin to act in an unacceptable way that could lead to punishment. Anxiety can also arise from the perception of actual threat in the external world. With further increases in the level of anxiety, the ego brings into play a series of defences to protect the organism from the destructive consequences of the developing tension. These 'Mechanisms of Ego Defence' are perhaps the most widely accepted parts of Freud's theories, partly because the mechanisms are stated in relatively precise terms, and because they seem to have the closest correspondence to personal experiences. These mechanisms were considerably refined and developed by Anna Freud (1936). Numerous mechanisms have been described by psychoanalysts, but only a few will be discussed here.

The most primitive, heedless and unselective of these mechanisms is *denial*, by which the presence of threat is completely dismissed from consciousness, and the person behaves as though it did not exist; this is possible mainly in young children with their incompletely developed egos, and in later years somewhat more sophisticated defences are marshalled in the presence of high anxiety, such as *repression*. Repression is a special type of denial, nearer to the surface of consciousness, and with a more refined and selective action. It includes forgetting threat or memories of threat, and keeps primitive impulses in the id and thereby out of awareness. Denial and repression are the most theoretically important and fundamental of the ego defences.

In *reaction formation*, anxiety-provoking impulses are so successfully suppressed that the person behaves in a way that is directly opposite to the direction of the impulses. Impulses are denied in the self and replaced by their opposite; strong sexual impulses, for example, may be reversed and appear as extreme opposition to the portrayal of explicit sexual themes in the arts.

Regression refers to the mechanism whereby the ego

returns to an earlier pattern of dealing with threat; for example, someone may burst into tears in response to a frustrating situation, or may revert to thumbsucking at times of stress.

Successful use of these mechanisms leads to their regular and consistent use as a pattern of coping behaviour, and they play an important role in the development of personality characteristics. However, excessive use of these mechanisms can result in the development of pathological symptoms.

The superego. The superego develops in turn from the ego, and is the structure that represents the standards and ethical values learned through contact with society in general, and with the parents in particular. It is closely connected to the ego, but retains considerable independence. The forces of the superego must be modified by the ego, since it is perfectionist and extremely unrealistic in the standards it sets. The ego tempers the standards to a more realistic level. Through the incorporation (or, more technically, the introjection) of phantasized parental values, the superego acquires internal means of differentiating good from evil; in short, it functions like an unrelenting conscience.

The three parts of the psychic structure are in continual conflict, with the id continually trying to obtain expression of its instinctual impulses, and the superego setting (often unattainable) moral standards. The ego has to keep these forces in the appropriate balance; sometimes it collaborates with the id, sometimes with the superego, depending on the current circumstances. Man's behaviour is partly a function of the ego's attempts to resolve the conflict between these impulses and partly a function of the ways in which conflicts and other experiences are dealt with at various stages of development.

Personality development
The libido is focused on different parts of the body at different stages of development. Most of these stages occur during the first five years of life, and if at any of these stages major conflicts are not resolved, the person may become 'fixated' at that

25

stage, producing a lasting effect on the person's subsequent character.

The first stage is the *oral stage*, the time when the infant is completely dependent upon others for its survival, and the libido expresses itself through sucking and chewing. Adult behaviours such as excessive gum-chewing and smoking, verbal aggression and other oral activities, are related to the person's experiences with early oral pleasures.

During the second year of life, the anus becomes the focus of libidinal drives; during this *anal stage*, toilet training is normally begun, and how this is done has profound effects on later development. The child learns that its bowel movements are ideal ways of controlling the parents' behaviour. Fixation at this stage can take several forms depending on the nature of the conflicts. Derivation of pleasure from the evacuation or retention of faeces may lead to sadism on the one hand, or obsessionality and meanness on the other.

Perhaps the most crucial stage is the one in which the libido is first centred in the genitalia – the *phallic stage*. The child has strong sexual urges towards the parent of the opposite sex. For boys sexual desire for the mother is accompanied by hatred and jealousy of the father, which in turn leads him to believe that the father wishes to punish him, and castrate him to prevent him achieving satisfaction of his incestuous desires. How the boy resolves this 'Oedipal' conflict has very important effects on later development. Successful resolution results in the boy modelling himself on his father, and inhibiting his incestuous yearnings. Unsuccessful resolution may lead to severe problems at later stages in life, particularly problems of sexual identification. Girls also have sexual desires towards their mothers, but this is inhibited when the girl realizes that she is without a penis, and that she may have been castrated. The original envy and aggression towards the father is gradually replaced by feelings of love towards him, and a desire for sexual relations with the father develops, but is kept under control by the ego. Successful conflict resolution at this stage is vital for normal character development in both sexes.

26

The *latency period* which follows is characterized by a diminution of sexual activity and a consolidation of the inhibition of sexual impulses. This lasts until adolescence, the onset of the *genital stage*, during which the person learns to direct sexual impulses towards appropriate objects and to enter into satisfying adult relationships.

Thus, classical psychoanalytic theory traces the roots of personality to conflicts within the three structures of the mind – the id, the ego and the superego; and to the resolution of conflicts at various stages of development. A person's behaviour at any moment will be governed by forces directly traceable to these factors.

There have been many developments within psychoanalysis, some of which have been major schisms resulting in entirely separate movements, such as Jung's 'individual psychology' and Adler's 'analytical psychology'. Most of these movements, although theoretically interesting, have not played a major role in contemporary thought on personality.

Major developments have taken place within the mainstream psychoanalytic fold and increasing attention has been paid to the development of the concept of the ego. Emphasis on the ego began with the work of Melanie Klein, and a modern version is referred to as *object-relations* theory. In object-relations, the biological roots of behaviour (i.e. the id) are given less prominence than in Freud's original writings; the influence that people have on each other receives a higher priority. In particular, the mother-child relationship plays a fundamental role in personality development, and forms the base upon which subsequent personal relationships are established. If the mother-child relationship is adequate, the person will develop an ability to form satisfactory relationships later in life; if not, conflicts and personal problems will arise. Guntrip (1971) has described the development and nature of object-relations theory and has emphasized its essential continuity with Freud's work.

Assessment related to psychoanalytic concepts

Freud did not develop an assessment technique to accompany his theory in the way that, for example, Eysenck has developed the EPI (see Ch. 4). Some 'projective' techniques have been designed to tap basic impulses and motives arising from the unconscious, and they have close links with psychoanalytic theory. (However, it should be remembered that many other techniques, including standard personality inventories and laboratory methods, have also been used to investigate psychoanalytic concepts.) The status of psychoanalytic theory is not, therefore, so critically dependent upon the status of projective devices as are other theories upon their associated assessment devices. Obviously Freudian ideas were the inspiration behind projective devices, but the influence is chiefly indirect.

Projective devices are designed to explore personality dynamics indirectly through the subject's responses to ambiguous and largely unstructured situations. Responses are not, therefore, taken at face value, but are valuable only in so far as they provide clues to underlying conflicts and defences. Subjects are not aware of the purpose of the test so that the ego defences are weak and indications of inner dynamics can begin to slip through, albeit disguised. The job of the tester is to interpret these indications. Three commonly used projective devices will be described.

The *Rorschach* test (named after its originator, a German psychiatrist) consists of a series of ten 'ink-blots', or complicated unstructured formless shapes, with one half the mirror-image of the other. The subject is presented with each card in turn, and he is asked to say what the blots could resemble. The tester notes down what the subject says, how long before he first responds, whether he uses the whole blot or just part of it, any repetitions or themes, and many other aspects of the responses. Common responses to some of the cards are, typically, 'a butterfly' or 'a bear-skin rug', and these are not suggestive of a significant underlying conflict. More abnormal responses such as seeing genitalia or 'death' in the blots do suggest serious unresolved conflicts.

The *Thematic Apperception Test* (TAT) consists of a series of pictures which vary in explicitness and detail, from a blank card and a few vague shadows at one extreme, to clear-cut scenes such as a photograph of a young boy sitting alone in a doorway, at the other. The subject is asked to examine each card in turn and to tell a story about the scene portrayed, including what preceded the scene, the current situation, and the likely outcome. Responses are timed and written down verbatim for later analysis. Interpretations are based mainly on any recurrent themes which may emerge in the stories; a subject with much repressed hostility may, for example, 'project' his hostility into the stories by frequent reference to fights, death or severe criticism.

One of the few projective techniques developed specifically in relation to psychoanalytic concepts is the *'Blacky'* test. Blacky is a puppy in a series of cartoon pictures, and the subject is presented with a series of cards depicting a variety of situations in which Blacky is involved with other dogs, such as his parents, siblings, and play-mates. The subject is required to explain what is happening in the cartoons, and the responses are alleged to provide clues to unconscious conflicts. In one cartoon, Blacky is seen with his tail about to be cut off, a picture designed to elicit Oedipal castration fears. Other cartoons assess sibling rivalry, parental rejection and other tension-producing experiences related to early childhood development. Obviously this test is used mainly with young children.

Literally hundreds of studies have been conducted into the reliability and validity of projective devices, and none of them fare very well; reliabilities tend to be of the order of ·2 at the most, and validity co-efficients are little better, if at all (Vernon, 1964). Correlations of this magnitude are too low for use as accurate assessments of an individual's unconscious processes or as devices to measure psychoanalytic concepts. In addition, Holmes (1974) has indicated that some of the basic assumptions of projective devices are not justified,

29

in that subjects can quite readily produce false and misleading responses, and can inhibit 'true' projections. He also concluded that projective devices are not necessarily tapping unconscious traits. However, some psychologists report that they can be useful in providing 'clinical hunches', to be followed up by more reliable and valid means.

Applications of psychoanalytic theory

There is almost no area of human behaviour which has not been examined or interpreted from a psychoanalytic viewpoint. Freudian ideas have guided theorizing in the disciplines of anthropology, mythology, philosophy, education, archaeology and criminology; they have been used in the interpretations of religious, mystical and political phenomena; and have put into a new perspective the study of history, art and literature. The processes of scientific discovery and creativity have been examined, and even economic crises have been said to relate to 'mass unconscious collusions'. But by far the most important application of psychoanalytic personality theory, and the one which arises most directly from its ideas, is the treatment of emotional problems.

The treatment of emotional problems

According to Freudian theory, overt behaviour is of therapeutic interest only in so far as it provides clues to underlying unconscious conflicts (see F1). All abnormal behaviour is determined by these conflicts; the target of therapy is not, therefore, to change the problem behaviour directly, but to examine the patient's defences and bring his conflicts to light. Once the conflicts can be identified, and the defences broken down, the problem can be seen in its true nature, and 'insight' is achieved. Insight alone is not, however, sufficient to resolve the emotional conflicts; the tension must be relieved by 'working through' the conflicts.

The psychoanalyst (as a therapist using Freudian techniques is called) serves as the recipient of the unconscious impulses that produced the tension. He functions as the object to whom the impulses are unconsciously directed, typically a parent. Hence understanding the relationship between patient and therapist is of profound importance. Since unconscious impulses can be motivated by either libido or aggression, some of the feelings within the therapeutic relationship will be sexual or aggressive. These emotional stages are called positive and negative 'transference', in that the unconscious impulses are transferred from the original object to the psychoanalyst. Many techniques are used to explore the patient's unconscious; the earlier techniques of 'free association' (in which the patient is required to express all his thoughts exactly as they enter his head) are now supplemented by observations of the patient's mannerisms, his posture, style of dress, social techniques in interacting with other people, or his hobbies and pastimes. No aspect of a person's behaviour is without significance.

It is almost impossible to judge the effectiveness of psychoanalytic treatment, despite its seminal influence on many forms of psychological treatment (apart from its importance as a treatment method in its own right), since so few studies have been conducted. There are many different reasons for this. Some psychoanalysts maintain that the effectiveness of therapy, from their own and others' experience, is so obvious that any attempt to demonstrate it in a 'scientific' manner is futile and irrelevant, and that to subject treatment procedures to a close analysis would be detrimental to the patient-therapist relationship. Attempts have been made to investigate effectiveness, but the experimental designs and methods of data analysis have often been so unsophisticated that the evidence is sufficiently ambiguous to permit considerable variation in interpretation.

Evaluation is rendered difficult in that, although psychoanalytic concepts are said to apply to the behaviour of everyone, only a few can benefit from psychoanalytic treatment. The most suitable patients are sophisticated, verbal and well-educated. Patients taken on for treatment are often carefully selected, and, if this selection procedure is used in a

31

study to examine the effectiveness of psychoanalysis (as presumably it should), it may render the study impracticable. A recent British study (Candy *et al.*, 1972) made a serious attempt to conduct a scientifically respectable investigation into the effectiveness of psychoanalytically-based therapy, but the project as planned was not feasible. Many of the patients referred to the study were rejected as not being 'eminently suitable for this form of treatment'; so many patients were rejected for this and other reasons, that the project finally had to be abandoned. This study illustrates some of the many complex problems that face the researcher into therapeutic outcome, and suggests why so few studies have been conducted.

A survey by Rachman (1971) concluded that the bulk of the available evidence would suggest that psychoanalytic treatment is of unproven value. Since Rachman's survey, new evidence has emerged from a very long-term study in the United States. The Menninger Project was set up to study psychoanalytic therapy in a clinical setting, and was planned and executed over a period of more than twenty years. The main finding was that the outcome of therapy was highly dependent upon the patient's characteristics, particularly how far his personal relationships were satisfying and adaptive and upon his ability to tolerate anxiety (Kernberg, 1974). Unfortunately, the project (which included only forty-two patients) did not compare psychoanalytic therapy with other forms of therapy, or with no formal therapy, and the results cannot, therefore, be regarded as providing unequivocal support for the effectiveness of psychoanalytic therapy. Rachman's conclusion need not, therefore, be revised in the light of the findings. However, the main importance of the Menninger Project lies in the demonstration that it *is* possible to conduct well-organized research in a clinical situation, paying due respect to the complexities of human behaviour and without prejudicing the patient-therapist relationship. The sophisticated methodology of this project augurs well for future outcome studies.

Evaluation of psychoanalytic personality theory

As we have seen, Freud's impact in a diversity of fields is enormous, but his influence on modern experimental psychology is quite minimal. This is remarkable considering the general importance of Freud's contributions to psychology, and the number of controversial and exciting ideas that his theories have generated. Forty years ago the experimental investigation of Freudian concepts was a major spur to research in psychology (summarized by Sears, 1944), but more recently Freudian concepts have been less intensively investigated.

There are many obstacles to the objective assessment of Freudian theories, not the least of which is connected with the language in which the theories are expressed. Freud was steeped in the science of the nineteenth century, and his writings show the heavy influence of contemporary hydro-mechanics, reflected, for example, in the frequent references to the build-up of tension which must be discharged through one outlet or another. It is often difficult to determine whether, in describing certain concepts, Freud intended his descriptions as metaphors and analogies to aid in understanding, or whether they were to be taken literally. The ambiguity and lack of precision frequently precludes establishing firm predictions by which to test the theory.

There are also some logical barriers to the objective investigation of Freudian concepts – the clearest and most often quoted example of these is the defence mechanism of reaction formation. Suppose that an interpretation of dream material leads one to hypothesize that a man's sexual orientation is homosexual, and suppose also that he frequently speaks out very forcefully against homosexuality. This would clearly be contrary to our expectations. However, if we refer to the mechanism of reaction formation, then his anti-homosexual statements may be seen as a reaction to his underlying strong homosexual impulses, and the conclusions based on the dream material are thereby supported. However, if he indulges in overt homosexual activities, then the conclusions would still be confirmed. When a hypothesis is stated so that even contrary results can be said to support it, the hypothesis is inherently untestable and any claims to scientific status must be queried (as indicated in Chapter 1). Similar criticisms apply to many other Freudian concepts such as regression and denial.

Despite these difficulties, some psychoanalysts have attempted to provide evidence for their claims. Much of this evidence, however, consists of case reports in which the author was also the therapist, and any conclusions are, therefore, liable to observer bias. In addition, much of the data consists of clinical material obtained from psychoanalytic therapy sessions and the need to provide independent evidence from outside the clinical setting (such as reports from patients' relatives, or actual behaviour) is rarely appreciated.

Kline (1972) has reviewed a large number of studies purporting to support Freudian theory. He applied very high standards of critical judgement to the methodologies of the studies and concluded that there is considerable scientific support for many of Freud's ideas and that the available evidence suggests that the validity of psychoanalytic personality theory has largely been demonstrated. Eysenck (1972) points out, however, that the studies reviewed by Kline indicate only that the existence of certain phenomena, such as repression, has been demonstrated, but that this does not necessarily provide support for Freudian theories alone. In other words, there may be several ways of interpreting the results, and to do so in Freudian terms only is unwarranted. For example, it has been frequently demonstrated that unpleasant ideas, wishes and memories are easily forgotten, and this is clearly in support of the Freudian concept of repression. But it is also equally in support of many other ways of interpreting the same results. Morris (1973) indicates that learning theory principles can provide an adequate account of repression, without resort to Freudian concepts such as ego and id. Such alternative explanations are not necessarily the 'true' ones, but may be equally plausible, and as long as there are alternatives, any statement to the effect that the

34

existence of Freudian concepts as such has been unequivocally demonstrated overstates the case.

Similarly, the evidence purported to demonstrate the existence of unconscious motivational phenomena has been critically reviewed by Brody (1972), and he concluded that there 'are no unequivocal demonstrations of unconscious phenomena. Research conducted under controlled conditions does not provide evidence for the kind of explanatory role assigned to the unconscious in traditional psychodynamic theories.' In view of the fundamental importance of unconscious factors in Freudian theory, this conclusion poses serious questions for those who embrace the psychoanalytic approach.

Recent developments such as object-relations theory provide a firmer and more systematic structure, which should permit the more explicit examination of Freudian theory; to date, however, the evidence in favour of the theory is sparse.

There is little in the details of psychoanalysis that originated with Freud. For example, Whyte (1962) has documented the long history of the concept of the Unconscious before Freud, and Stengel (1973) has described the profound influence of Hughlings Jackson, a nineteenth-century American neurologist, on psychoanalytic thought. However, it was a major achievement and a creative intellectual feat to gather together the many different strands and combine them into a complete and coherent system.

One of Freud's notable contributions to the study of personality was his influential insistence that man's behaviour is caused, and does not happen simply at random or because of chance factors. For Freud, all behaviour was determined and even what appeared at first sight to be simply a chance happening could ultimately be shown to have originated from the Unconscious, and was potentially explainable. One must assume man's behaviour is caused before it can be examined in a scientific framework and Freud played a major role in propagating this idea.

Before Freud, conventional wisdom maintained that man was a logical, rational being who behaved in a purposeful manner, always aware of the reasons for his behaviour. By emphasizing unconscious motivational influences, Freud contributed greatly to the destruction of this belief and henceforth man has viewed himself in a more humble perspective.

Freud's theory of personality has been criticized largely because it does not conform to the tenets of contemporary philosophies of science. On the other hand, it has been argued that if philosophies of science cannot accommodate the grandeur and vision of psychoanalytic theory, so much the worse for philosophies of science.

3
Interpersonal theories

Although there are a number of important differences between the two main proponents of interpersonal theories, Carl Rogers and George Kelly, they share a similar view of man. They paint a picture of human behaviour, more noble, rational and active than that of other theorists. A leading supporter of Kelly's views has expressed the contrast in the following terms:

> Thus psychoanalytic theories seem to suggest that man is basically a battlefield. He is a dark cellar in which a well-bred spinster lady and a sex-crazed monkey are forever engaged in mortal combat, the struggle being refereed by a rather nervous bank clerk. Alternatively, learning theory seems to suggest that man is basically a ping-pong ball with a memory. Along these lines some types of information theory hint at the idea that man is basically a digital computer constructed by someone who had run out of insulating tape. (Bannister, 1966, p. 21)

He suggests that few psychologists would accept such limited descriptions of themselves.

Rogers and Kelly are both concerned with the whole person. There is no attempt either to appeal to underlying physiological or neurological mechanisms to explain behaviour or to fragment the individual's behaviour into various functional systems, viz. perception, memory, motivation. The individual reacts consciously to his own interpretation of the world about him and tests this subjective view against objective reality. Thus man is not driven by inner urges or controlled by his environment but is at all times actively trying to make sense of his experience.

The personality theories of Rogers and Kelly both originated from their work in the fields of counselling and psychotherapy. There is no attempt to delve into the hidden meanings of patients' behaviour, as in psychoanalysis. The individual is regarded as the best expert on himself and his statements about himself constitute the raw-material of therapy. Their preferred assessment techniques (see later) are also more open-ended and allow for greater self-expression than questionnaire methods. However, they also allow for greater objectivity and a more systematic approach than projective techniques.

In both theories we find a strong emphasis on experimentation. It is only when clinical observation has been confirmed by carefully controlled research that it can form a sound theoretical base. Kelly and Rogers see their theories of personality as having a limited life. Their main function is to stimulate research and thus contribute to their own demise when the discovery of new evidence allows more adequate theories to be constructed.

Rogers' 'self' theory

The major element in Rogers' theory is the concept of 'self'. There have been many 'self-theorists' in psychology and each has contributed to our understanding of this important area. Rogers became aware of the importance of 'self' through his experience with clients in psychotherapy. Patients tended to describe their experience by reference to a 'self' and their problems often seemed to arise out of inconsistencies in their view of themselves – 'I don't understand how I could act that way' – 'I feel different from how I used to feel.'

The 'self' in Rogers' theory is the organized pattern of perceptions, feelings, attitudes and values which the individual believes are uniquely his own. They are the characteristics which define 'I' and 'me'. Thus the self is the central component of the total experience of the individual. An important additional concept in Rogers' theory is the 'ideal self' which is the person as he would like himself to be. The well-adjusted individual is regarded as having a fairly close correspondence between the self and the ideal-self.

The major activity of the total individual (which Rogers calls the 'organism') involves reaction with the field of experience in a way which satisfies its needs. The basic motive of the organism is to actualize, maintain and enhance itself. The tendency towards 'self-actualization' gives meaning and direction to human activity. The trend is towards greater differentiation and integration, the individual becomes more independent and has more freedom of self-expression. Rogers is suggesting that the 'self' and the 'organism' are both involved in initiating and controlling behaviour. As long as the two systems are working in harmony and there is consistency between the self and the experience of the organism the individual remains 'congruent'. Where he experiences a discrepancy between his perceived self and his actual experience the individual will be in state of 'incongruence' which results in tension and maladjustment.

Most experiences are consciously perceived by the individual but Rogers believes that experiences which are too threatening to the individual's self-concept may be denied awareness. When the person is aware of a state of incongruence he may change his behaviour to maintain his self-concept or modify his self-concept in the light of his inconsistent behaviour. Where inconsistent behaviour is denied awareness, the individual experiences anxiety and this may lead to the operation of defence mechanisms which act to maintain the integrity of the self. The defence mechanisms recognized by Rogers are similar to those described by Freud (see pp. 24–5).

39

The initial statement of Rogers' theory (Rogers, 1951) left a number of questions unanswered concerning the gulf between the experience of the organism and the self and the consequent need for defence mechanisms to protect the self-concept. In order to fill this gap Rogers (1959) suggested the concept of the 'need for positive regard'. This need for positive regard which may take various forms such as the love, approval, sympathy, respect of other people, is thought to be present in everyone. It is seen most clearly in the young child's need for the love and approval of its parents. The child's self-concept will be shaped by the parents' use of conditional positive regard. If the child's actual experiences conflict with the self-concept 'demanded' by his parents as a condition of love, then defence mechanisms may have to be employed to prevent the experiences from entering awareness, e.g. if the parents value academic success, the child is likely to develop a self-concept which emphasizes high academic achievement; failure to match up to his own high standards may be dealt with by employing defence mechanisms. Thus the origins of defence mechanisms are found in the individual's attempts to retain love. The need for positive regard may become so powerful that it overshadows the more basic biological needs of the organism. Thus in Rogers' theory examples of martyrdom and self-sacrifice do not appear so contradictory as they do in other personality theories.

The neurotic individual is forced to make excessive use of defence mechanisms because his actual experience is too far out of step with his self-concept. The result is anxiety and rigid maintenance of the self against all inconsistent experience. As more and more experiences are denied awareness, the self loses touch with reality and the individual becomes increasingly maladjusted. Thus Rogers views psychological disorders as essentially part of a continuum in which acute psychotic breakdowns result from behaviours inconsistent with the self breaking through the defensive barriers into awareness.

Although it is the discrepancy between self and experience which is central to Rogers' theory of psychological disorder, this aspect has been comparatively neglected by researchers. The major interest has centred on the discrepancy between self- and ideal-self-concepts and measures of these have been widely employed in research into maladjustment and response to psychotherapy.

Assessment
A technique for investigating the self-concept, known as Q-methodology, was developed by Stephenson (1953). It is possible to use Stephenson's Q-technique to describe the self-concept without employing the full methodology, and this is how it has normally been used in research into Rogers' personality theory and the effects of psychotherapy.

The investigator takes a large number of descriptive statements which the subject has to sort into a number of different piles, ranging from 'least characteristic' to 'most characteristic' of himself. The subject is instructed to place a standard number of cards in each pile so that most of the items will be bunched in the middle of the continuum with few at the extremes.

In addition to obtaining a Q-sort in terms of a subject's self-concept, the investigator may ask the subject to re-sort the same items within a new frame of reference, e.g. ideal-self, myself as a child, myself as my wife sees me. It is also possible for the subject to carry out the Q-sort as he believes it would apply to another person, e.g. husband, wife, father or mother. Where different Q-sorts have been obtained such as for 'self' and 'ideal-self', a correlation coefficient may be calculated which gives a measure of the discrepancy between the two concepts. The greater the discrepancy between 'self' and 'ideal-self' the greater the maladjustment. If the Q-sort is repeated at various points in the course of psychotherapy it is possible to obtain a measure of change. The discrepancy between self and ideal-self would be expected to become smaller as a result of successful therapy and it would be possible to determine whether the overall change resulted from changes in the self-concept, the ideal-self-concept, or both.

This rather simple view of maladjustment has been criticized and a number of studies suggest that the expected changes in self- and ideal-self-concepts do not always occur. A study by Truax *et al.* (1968) found agreement between the results of Q-sorts and other psychological test measures of adjustment and change in groups of neurotics and delinquents. However, Garfield *et al.* (1971) examined the relationship between eight different criteria of outcome in psychotherapy, including Q-sort and ratings by clients, therapists and supervisors, and found little agreement between the different measures. In addition, there is some evidence that readministration alone may produce changes in the Q-sort in the absence of psychotherapy. One reason for the difficulty in establishing clear evidence of change by means of Q-sorts may be that the client tries to appear more well-adjusted than is actually the case. This suggests that the use of the discrepancy between self and ideal-self may result in a number of seriously maladjusted individuals being classed as well-adjusted.

The Q-sort is a structured technique in that the items are provided by the experimenter and the subject must sort them into predetermined categories. However, it does allow for considerable flexibility in its application to the individual case. The Q-sort accepts what an individual says about himself at face value. This is in line with Rogers' basic position that an individual is the best source of information about his own feelings, beliefs or attitudes. Unfortunately, even if we accept the validity of the technique, bearing in mind the evidence that individuals may often make responses which they believe to be expected by the experimenter or which they regard as socially desirable, then we are still left with doubts about its test-retest reliability. Although the Q-sort may be useful as a research technique for investigating the self-concept with groups of subjects, its clinical use with individuals, particularly as a measure of change, appears to rest on insecure foundations because of its low reliability.

Applications
In many respects Rogers' theory is a theory of psychotherapy and most of the recent research has been concerned with

42

aspects of the therapeutic relationship. Psychotherapy is regarded as a specialized example of constructive interpersonal relationships and Rogers has attempted to set out clearly the conditions which he believes are necessary to produce personality change.

In his initial presentation of his therapeutic method Rogers (1940) laid great stress on the therapist being non-directive. The main function of the therapist was to reflect accurately the client's emotions so that the client could recognize and clarify his own feelings. There was no attempt to deliver interpretations or to impose the therapist's views upon the client. Rogers subsequently became convinced that his principle of non-direction was being misinterpreted by some therapists who were adopting a position of disinterest in the client. To counter this he suggested that therapy should be 'client-centred'. The therapist was expected to explore with the client his total field of experience in order to understand the client's own view of himself.

In a theoretical article Rogers (1957) emphasized the six essential conditions for constructive personality change which he believed would be found in all effective treatment settings with all types of clients.

1 Two persons are in psychological contact.
2 The first, whom we shall term the client, is in a state of incongruence, being vulnerable or anxious.
3 The second person, whom we shall term the therapist, is congruent or integrated in the relationship.
4 The therapist experiences unconditional positive regard from the client.
5 The therapist experiences an empathic understanding of the client's internal frame of reference and endeavours to communicate this experience to the client.
6 The communication to the client of the therapist's empathic understanding and unconditional positive regard is to a minimal degree achieved. (p. 96)

As the direction of constructive change is assumed to be always towards greater self-actualization, the therapist's responsibility is to provide the necessary conditions of congruence (genuineness), unconditional positive regard (warmth), and empathy. He does not have to control or direct the therapeutic process. When the necessary conditions are provided and maintained over a period, constructive personality change will occur.

A number of studies by Truax and Carkhuff (1967) have shown that the three crucial ingredients of empathy, warmth and genuineness are indeed provided by effective therapists who work from a variety of different theoretical bases. The major research technique has involved the rating of short sections of interview tape-recordings by trained raters on carefully specified scales. The inter-rater reliabilities appear for the most part to be adequate though in some instances, particularly in the case of 'genuineness' ratings, they may be less acceptable. There is little evidence on the validity of the scales. The three therapeutic conditions are generally found to be positively correlated though in some studies one condition has been found to be negatively correlated with the other two, while in other studies no relationship has been established. This leaves the relationship among the therapeutic conditions rather uncertain.

Truax and Carkhuff (1967) have reviewed a large number of studies which support the view that the presence of high levels of therapeutic conditions is associated with a positive therapeutic outcome. The presence of low levels, however, does not necessarily lead to a deterioration in the client's adjustment. There have been a number of inconsistent results and some studies have failed to demonstrate significant change following treatment (Garfield et al., 1971). As client-centred therapy is normally of short duration, there has been an important assumption that positive change continues after the termination of therapy. Fiske and Goodman (1965) used a variety of measures such as Q-sort, TAT and MMPI and found some additional improvement eighteen months after therapy, but other clients had relapsed, and there was no significant improvement for the group as a whole during the post-therapy period.

44

Although adequate levels of empathy, genuineness and warmth are probably necessary conditions for effective personality change, the importance of additional factors seems to be indicated by some of the inconsistent findings mentioned above. It has been suggested that the level of 'self-exploration' which requires the client to explore his feelings, values, relationships and fears, may be an important determinant of successful outcome. Truax and Carkhuff are able to provide some evidence that ratings of high levels of therapeutic conditions are associated with greater self-exploration and personality change. There is also evidence that Rogers applies selective verbal and non-verbal reinforcement to the client's statements, which has the effect of promoting client self-exploration and re-evaluation of his own behaviour (Truax, 1966). This finding is contrary to Rogers' earlier requirement that therapy should be non-directive.

In reviewing the work of Rogers and his associates into the psychotherapeutic process, one is left with the feeling that more questions have been raised than answered. Nevertheless, a promising beginning has been made into the scientific investigation of this fascinating but complex area and for this considerable credit must be accorded to Rogers and his fellow workers.

Evaluation
The type of self-theory proposed by Rogers regards the 'self' as the most important aspect of the individual. This view has not always received wide support. Epstein (1973) considers that no one has succeeded in adequately defining the self-concept and most definitions are circular or lack observable reference points. However, he considers that:

Although there is disagreement about the value of the self-concept as an explanatory concept, there can be no argument but that the subjective feeling state of having a self is an important empirical phenomenon that warrants study in its own right. (p. 405)

The chief merit of Rogers' approach to the self-concept is

his emphasis on assessment and research. The theory is only elaborated at a pace consistent with the growth of experimental data. Also Rogers recognizes that other approaches to the study of human behaviour are possible and that they may make valuable contributions to self-theory.

One criticism of Rogers' theory is that it places too much reliance on conscious cognitive processes to the relative neglect of unconscious aspects of behaviour. Self-reports are notoriously unreliable, not only because the individual is influenced by the expectancies of the investigator or social desirability effects, but also because a person rarely knows the whole truth about himself. Of course, the vantage point of the observer is equally subjective and open to distortion, so that despite its basic unreliability, self-report may often be the best source of information about an individual (see also p. 123).

Personality development is treated rather superficially by Rogers and the mechanisms of personality growth and change are not stated with sufficient clarity. The emphasis has been on the conditions favourable to growth and change and the direction of change is assumed to be towards greater self-actualization. In their more recent writings Rogers and his associates acknowledge the importance of other factors, such as selective reinforcement, in determining the pattern of personality change.

Rogers' theory and his approach to counselling have had a considerable influence on psychotherapists, teachers, the clergy and business executives. His major achievement has been to bring the practice of psychotherapy into the realm of scientific study and to stimulate research into the more complex aspects of human behaviour.

Kelly's personal construct theory

Personal construct theory (PCT), which Kelly regards as a cognitively orientated comprehensive theory of behaviour, grew out of the observation that all psychological theories contain two views of man. The first and more explicit aspect of the theory views man as a collection of conditioned responses or the victim of his childhood fantasies and so on, while the second, implicit, theory is of man the behavioural scientist who creates theories, tests hypotheses and attempts to control his environment. Kelly argues that few psychologists would accept the former view of *themselves*. Their conception of themselves as scientists is altogether more noble and dignified than that contained in the theories which they use to understand the behaviour of their (human) subjects. Thus he is asserting that psychological theory should be 'reflexive'; in that it should account for the behaviour of the person who creates the theory as well as accounting for the behaviour of his subjects.

Man is regarded as a 'scientist' who is actively trying to make sense of his environment. He constructs theories, tests predictions and weighs the experimental evidence. The basic philosophical assumption in Kelly's theory is that all events are subject to *alternative constructions*. There is no absolute truth or objective reality but only ways of interpreting events (constructs), which are more or less useful in advancing our understanding and ability to predict future events. Thus the essence of Kelly's theory is contained in a single fundamental postulate: 'A person's processes are psychologically channellized by the ways in which he anticipates events.' (Kelly, 1955)

In PCT we find little mention of motivation (or learning or perception or emotion); the person is considered to be essentially active. He is at all times trying to enhance his understanding of events and he does not need to be pulled or prodded into motion.

The person understands the world in terms of 'constructs' (concepts) that have predictive utility for him. Thus a construct is more than a mere label; it is a way of predicting future events. If we apply the construct 'trustworthy' to an acquaintance then we are making a prediction about how we expect him to behave in a position of 'trust'. The usefulness of a construct is determined by the accuracy of the predictions we make from it. All constructs involve a contrast between

47

two opposite poles (e.g. hard-soft, tall-short) and constructs arise when an individual 'construes' two persons or objects as sharing a common characteristic which differentiates them from a third. It is not possible to apply a particular construct to all situations since some things will lie outside its 'range of convenience'. Equally, there will be other circumstances in which the construct is extremely useful and this constitutes its 'focus of convenience'.

Although most constructs can be named, Kelly recognizes that some important constructs may be non-verbal and unavailable to conscious awareness. If only one pole of a construct is verbalized, the other pole is said to be 'submerged'. Thus, if an individual commonly uses the term 'clever' to describe people, he is implying the existence of the opposite pole 'stupid' though he may not use it. Constructs may be organized in a hierarchical fashion such that a wide-ranging construct like 'good-bad' is in a 'superordinate' position relative to a number of 'subordinate' constructs, which have narrower ranges of convenience. The more stable superordinate constructs constitute the 'core constructs' which give the individual a sense of identity and continuity. These core constructs can be contrasted with 'peripheral' constructs which are usually subordinate and can be altered without seriously disturbing the core constructs. In many ways the core constructs, particularly those which the individual habitually uses to evaluate his own behaviour, can be considered analogous to Rogers' self-concept.

Those constructs which are flexible in that they are able to assimilate new elements into their range of convenience are considered 'permeable', whereas others are relatively 'impermeable' in that they rarely allow new elements to be admitted. Those constructs which are closely correlated with others are called 'tight' constructs and they lead to unvarying predictions (e.g. this is a 'table' therefore it is 'solid'). When constructs are 'loose' it is possible to make a greater variety of predictions in similar circumstances. In general, construct systems should not be too 'tightly' organized or they will prove too restrictive; nor should they be too 'loosely' organized in which case the accuracy of the predictions made from them becomes difficult to assess. Individuals may differ not only in the constructs which they use but also in the ways in which their construct systems are organized so that each individual has a unique personal construct system. The structure and organization of an individual's construct system constitutes his personality.

The concepts of 'anxiety', 'threat', 'hostility' and 'aggression' are important in Kelly's theory though he provides novel definitions for these terms. 'Anxiety is the recognition that the events with which one is confronted lie outside the range of convenience of one's construct system.' It is not the invalidation of one's constructs which produces anxiety but the knowledge that one's constructs are inadequate for understanding the situation. 'Threat is the awareness of imminent comprehensive change in one's core structure.' As the individual seeks to elaborate his construct system by engaging in new activities and entering new situations he exposes himself to anxiety and threat. A person may react to anxiety and threat by tightening and constricting his construct system or by broadening and loosening it. If the process goes too far in either direction then psychological disorder may result.

Kelly's approach to emotions is deliberately psychological but in order to achieve this position he is forced to ignore a wealth of knowledge from the field of physiology; furthermore some of the definitions appear to fly in the face of commonsense. Bannister and Mair (1968, p. 33) state that 'Within this scheme "emotions" lose much of their mystery'; it can be argued that they also lose most of their meaning.

Assessment
The Role Construct Repertory Test (Rep. Test) devised by Kelly is closely related to his basic theory. It is a method of establishing the basic constructs which a person uses and the inter-relationship between them. The subject is given a list of roles (e.g. teacher, close friend, employer) and asked to supply the name of someone who fits the role. He is then presented with three cards, each bearing one of the names

49

he has supplied, and he is asked to say in what way two of them are alike but different from the third. This procedure establishes the similarity and contrast poles of the construct (see p. 48). The process is repeated with different cards until a sufficient number of constructs has been elicited. Various modifications of the method of eliciting constructs are possible, but one of the most useful is the 'Self-Identification Form' in which the subject's own name is included in each presentation of the group of three cards. The constructs elicited from the Rep. Test can be analysed in terms of their content or in terms of how the subject applies his constructs to particular individuals or situations.

The original Rep. Test was intended for individual use by the clinician but a method developed from it, called the Repertory Grid (Rep. Grid), has become a major research technique in personality and psychopathology. It allows the investigator to use many different types of elements (e.g. role titles or names). Constructs may be elicited or provided by the investigator, and a variety of sophisticated scoring systems are available. Using either elicited or provided constructs, the subject may be required to assign each role title (i.e. name of an important person) to one or other end of the construct pole, or to rank-order them in relation to the construct, e.g. to assign the elements 'mother', 'employer', 'girlfriend' to one pole of the construct 'honest-dishonest', or to rank them in terms of their 'honesty'. When the first method is used a 'matching score' is obtained which is a measure of the degree of association between the constructs. A similar 'relationship score' can be calculated using the rank-ordering method. It is assumed that statistical measures of association reflect the psychological relationships of the construct system.

Other important assumptions underlying Rep. Grid methods are that the persons or roles chosen are representative of important figures in the subject's life, that the sortings chosen by the investigator are representative of those that the subject faces in his daily decisions and finally that the labels which are attached to the constructs mean much the same to the subject as they do to the investigator. The evidence supporting these assumptions is, unfortunately, rather sparse.

The problems of validity and reliability have received relatively little attention. Bannister and Mair (1968) argue that in their traditional form these concepts are not strictly relevant, as the Rep. Grid is a methodology rather than a standardized test. This position has not gone unchallenged. A study by Gathercole *et al.* (1970) investigated 'parallel-form' reliability (different persons in the same role titles) and 'test-retest' reliability of types of Rep. Grid in general use. They concluded that generalizations about individuals based on single grids, particularly if elicited constructs are employed, should be made with extreme caution since the results are likely to be unreliable. In their extensive review of Rep. Grid technique and its applications Bannister and Mair (1968) suggest that:

> There is some danger that grid method may suffer the fate of the Rorschach. It may surround itself with a mass of loose folk-lore and semi-norms which enable the psychologist to practise casual speculation disguised as systematic instrumental investigation (p. 201).

There is unfortunately some evidence that this timely warning has not been heard.

Applications
Psychiatric diagnosis. Considerable research effort has been prompted by Kelly's speculation that the odd thought processes sometimes found in schizophrenics are the result of abnormally 'loose' construct systems. Bannister and Fransella (1967) published a 'Grid Test of Schizophrenic Thought Disorder' in which subjects were required to rank-order eight photographs of unknown individuals on six constructs. The test is then repeated using the same photographs and the same constructs and two scores 'intensity' and 'consistency' are calculated. Intensity is a measure of the degree of association between the constructs, i.e. how far individuals judged to be 'kind' are also judged to be 'sincere'. The consistency

51

score measures the stability of the rankings between the two administrations and is in effect a measure of short term test-retest reliability.

This test has been widely used to diagnose schizophrenic thought disorder, though its utility for this purpose has not been firmly established. The original standardization used subjects judged clearly thought disordered or non-thought disordered by three independent raters; thus it may not be appropriate for use with patients showing less clear-cut thought disorder. In an attempt to provide further information on the test, Bannister *et al.* (1971) administered it to over 300 unselected psychiatric patients drawn from three different hospitals and examined the results in relation to clinically diagnosed thought disorder. Their conclusion is that the relationship is satisfactory. However, as Poole (1973) has shown, the rate of mis-classification (calling non-thought disordered subjects thought disordered and vice versa) in Bannister *et al.*'s study is too high to justify using the test with individual patients.

The concept of 'loose construing' as an explanation for schizophrenic thought disorder has often been criticized, since it appears to be consistency and not intensity (i.e. the tightness or looseness of the construct system) that distinguishes thought disordered schizophrenics from other groups. Radley (1974) has reviewed the experimental evidence and concluded that there is no justification for differentiating between 'loose' and 'cognitively complex' individuals on the basis of measures of 'intensity'.

Psychotherapy. Kelly considers the purpose of psychotherapy to be the modification of the construct system. The major technique described by Kelly, 'Fixed Role Therapy' (FRT), has been comparatively little investigated, and controlled trials to evaluate its effectiveness have not been conducted. Nevertheless, it has shown some early promise. An introduction to FRT which contains an illustrative case description will be found in Bonarius (1970).

The first step in FRT is to establish the most important

52

constructs in the client's personal construct system by means of self-description or the Rep. Test. The therapist then draws up a role sketch which the client is required to act in his daily life. The main constructs of the role must be completely different from those of the client, to avoid the danger of his using his own constructs to play the role and thereby failing to experience a new approach to his life situation. The role must also be a 'natural' one since the client is not supposed to tell other people that he is acting a role. The therapist uses the therapy sessions to check the client's progress in playing the role and to rehearse aspects of the role with the client. Finally, the role sketch is returned and the client discusses with the therapist what he has learned from the experience.

The process of change is considered to be the experimentation with new constructs; a person who comes to construe things differently and to behave differently will be a 'different person'. However, it seems unnecessary to evoke the elaborate structure of PCT to explain any changes which may occur in FRT, especially since the technique appears to have many similarities with the method of 'modelling' derived from social learning theory (see Ch. 7). It is suggested that social reinforcement by the therapist and others who interact with the client in his 'new role' is an equally plausible mechanism of change. This explanation may be preferable because it is more parsimonious and is derived from a theory (social learning theory) with a much firmer base and wider explanatory range of convenience than PCT.

Evaluation

PCT constitutes a brave and imaginative attempt to create a comprehensive, cognitive theory of personality. Where it fails it does so because it ignores or trivializes important aspects of human behaviour. The theory fails to deal adequately with emotion; Bruner says:

I rather suspect that when some people get angry or inspired or in love, they couldn't care less about their 'system

53

as a whole'. One gets the impression that the author is, in his personality theory, overreacting against a generation of irrationalism. (1956) p. 356.

Situational factors are also neglected and PCT appears to place man in an 'empty world'.

Although Kelly escapes the paradox of 'reflexivity' (see p. 47) his theory is paradoxical in other respects. Mischel (1964) points out that since Kelly assumes that there is no 'objective reality', but only a person's subjective interpretation of reality, it is impossible for constructs to be objectively invalidated. He claims that constructs are used to classify and not to predict so that human behaviour with regard to the future is essentially prescriptive rather than predictive. Scientific predictions have no power to influence events though a man's predictions may, e.g. 'I will marry a certain type of girl.' A further paradox is encountered in the therapy situation since it appears that we may be faced with a therapist purporting to know the client's construct system better than he does himself. Perhaps the most critical issue for PCT is how far the predictions derived from the theory can be made from simpler theories which have a greater explanatory scope.

In creating his model of 'Man the Chess Grand Master' Kelly has drawn attention to important areas of personality long neglected by researchers. That future theorists will have to pay greater attention to cognitive variables is certain; however, the theories they adopt are likely to have a narrower range of convenience than PCT.

A comparison of Rogers and Kelly
The interpersonal theories of Rogers and Kelly draw attention to important areas in personality, viz. the 'self' and 'cognitive processes'. The price of this emphasis is a reduction in comprehensiveness and a relative neglect of important areas such as unconscious processes, learning and biological factors. Both theories are essentially clinical theories and they share some similarities with the psychoanalytic theory discussed in Chapter 2. An important difference from psychoanalytic

theory arises because the theories are framed in a way which permits scientific investigation. Thus they are capable of invalidation and Rogers and Kelly have shown a greater willingness to revise their theories in the light of new evidence than most other theorists.

The assessment measures, Q-sort and Rep. Test, favoured by Rogers and Kelly have shown some promise though there remain problems concerning reliability and validity. In the clinical field, in particular, the value of these assessment devices has still to be conclusively demonstrated. The methods of psychotherapy associated with the theories require further empirical evidence of their effectiveness, in view of a number of inconsistent findings. The process factors in psychotherapy, derived from the theories, have not been well supported and it is suggested that social-learning principles may make an important contribution to our understanding of the mechanisms involved.

The dangers of erecting a theory of personality on a clinical base are not simply those of limited perspective; they also contain the more important problem of distortion due to the special effects of stress and the particular characteristics of the psychotherapeutic relationship. In the case of Kelly's theory the unrepresentativeness of his subjects may be even more pronounced, as Foulds (1973) has pointed out:

If general psychology has often placed disastrous reliance on university students, psychopathology – or psychopathology as understood by many clinical psychologists – is placing disastrous reliance on the clientele of university clinics.

4
Type and trait theories

'A "trait" is a determining tendency or a predisposition to respond.' (Hall and Lindzey, 1957, p. 263) In some respects everyone is a trait theorist, in that we note consistencies in other people's behaviour and we label them accordingly as aggressive, lazy, shy or by one of the other 18,000 words which exist in English to describe human behaviour. The number of trait-type descriptions available in ordinary language is too large to be employed in personality research and some method has to be found of reducing the number to more manageable proportions. The majority of investigators in recent years have used a specially designed mathematical method called factor analysis.

The personality theories of Cattell and Eysenck which will be considered in more detail in this chapter both make extensive use of factor analytic techniques. The differences in emphasis between the theories of Eysenck and Cattell result in part from theoretical differences in the methods of factor analysis used by the two researchers. In order to understand these differences it is necessary to present an outline of the factor analytic methods used in personality research before proceeding to a description of the personality theories.

The technique of factor analysis

If we take a number of tests which measure some aspect of an individual such as his body-build, the speed with which he solves simple arithmetic problems, or his heart-rate fluctuations under emotional stress, then the degree of relationship between the various measures can be expressed by a correlation coefficient (see A9). If a coefficient of $+0.85$ is obtained between two tests then we expect similar results, i.e. high scores on one would be associated with high scores on the other test and they could be said to be measuring the same thing. Coefficients close to zero such as $+0.1$ would suggest that there is little relation between the two tests and that they are not measuring the same factor.

A typical factor analytic investigation of personality would involve the collection of a wide variety of measures from a large number of subjects. The investigator would then calculate the correlation coefficients between all the different measures. The next step is to arrange all the correlation coefficients in a table called a correlation matrix. It is difficult to make sense of the vast array of numbers contained in the matrix so some method of simplifying the information has to be employed. Factor analysis enables us to determine the minimum number of dimensions or factors which will summarize the data in the correlation matrix by noting the 'grouping together' of correlations. It is often only necessary to extract two or three factors to account for most of the information contained in the matrix. How far a particular factor is related to a particular test can be determined by the correlation between them, and this relationship is known as the test's *'loading'* on the factor. Usually, a test will load on a small number of different factors.

The apparently precise mathematical nature of factor analysis, which is one of its chief attractions as a research tool, at the same time constitutes one of its major pitfalls. It is attractive because it enables complex data to be quantified and reduced to a more manageable form. Its chief danger is that its very numerical precision may obscure the number of intuitive judgements which have entered into the research process and the subsequent analysis of the data. First, the researcher makes a decision about the tests he will employ and the type of subjects he will study. If meaningless rubbish is collected as data then rubbish will emerge from the analysis, though this fact may be obscured by the neatness of the mathematical computation and the precision with which the figures are stated.

Even the process of factor analysis itself contains a number of arbitrary decisions about the way in which the data will be handled. An important decision which the personality researcher has to make is whether to use *orthogonal* or *oblique* methods of factor analysis. When an orthogonal factor analysis is undertaken the various factors which emerge should be independent of each other (i.e. uncorrelated). Thus knowledge of an individual's score on a particular factor-dimension will tell us nothing about his score on another factor-dimension. Oblique factor analysis, on the other hand, accepts that the factors may be correlated to some degree and are not, therefore, completely independent. The correlation coefficients of the factors can be arranged in a new correlation matrix and a second factor analysis carried out. The factors produced by this second analysis are called second-order factors and they in a sense represent groupings of the original (first-order) factors. We often describe the first-order factors as *'traits'* and the second-order factors as *'types'*.

The fact that Eysenck prefers to use orthogonal analysis and hence obtains a small number of independent factors, makes his theory essentially a 'type' theory of personality. Cattell's preference for oblique analysis means that he arrives at a larger number of correlated factors and his theory is, therefore, basically a 'trait' theory of personality. However, the difference is not purely a mathematical one in that each believes that his chosen level is more in keeping with psychological reality. However, these differences may be more apparent than real, since Cattell's chief second-order factors seem to correspond closely with the two major 'types' found by Eysenck.

When we obtain a factor-dimension from a factor-analysis we have to decide what to call it. This decision is usually based upon an examination of the tests which load on the factor, to see what the tests have in common. This is largely a subjective judgement and other workers may not always agree with the labels chosen. In order to minimize this difficulty Cattell prefers to refer to his trait-dimensions by their number in a classification index so that no unwarranted assumptions are made about what the factor actually represents or measures.

It is tempting to believe that the factor-dimensions which the researcher discovers in his factor analysis are more than mere mathematical abstractions; and that one may assume that the factor structure in some way describes or represents the psychological structure of intelligence or personality or whatever concept is being investigated. Early factor theorists certainly believed this but recently more caution has been observed. As long as we remember to treat the traits or types isolated by factor analysis as convenient abstractions, as we do the concepts of intelligence, or gravitation, and not as concrete structures, then they may prove extremely useful.

The remainder of this chapter will describe the theories of Cattell and Eysenck and the methods of assessment associated with their theories.

Cattell's trait theory

In formulating his theory of personality Cattell assumes that there are natural elements of personality which can be discovered by the technique of multivariate analysis. In this procedure a large number of naturally occurring aspects of human behaviour are sampled in an attempt to unravel complex interrelationships and uncover the basic building blocks of personality. Cattell believes that we all have the same traits but to different degrees, and in this way no two individuals are exactly alike.

Cattell began his researches by taking trait names used in everyday language to describe human behaviour together with more technical descriptions drawn from psychology and psychiatry. He then rated individuals on the trait descriptions and inspection of the ratings suggested that about forty basic dimensions were involved. When individuals were rated on these dimensions and the ratings were factor analysed it was found that fifteen factors could account for all the terms used to describe people.

The next step in the development of Cattell's theory involved the construction of questionnaire measures of personality. Very large numbers of questionnaire items were drawn up and analysed. Of the sixteen factors which emerged twelve showed close similarity to those already obtained from ratings. Thus four quite new factors emerged from questionnaires and three of the earlier ones did not reappear.

The third stage in the project was to administer a variety of objective behavioural tests to a large group of individuals and to factor analyse the results. These objective tests involve observation of an individual in a structured situation so that predictions can be made about his behaviour in other situations. In all, twenty-one factors were found and when these were compared with the second-order factors obtained from the questionnaire data a number were found to correspond. Unfortunately for Cattell's theory there does not seem to be a common factor structure underlying the data drawn from the three separate sources of information which have been examined.

The sixteen first-order factors or 'primary' factors which make up the Sixteen Personality Factor Questionnaire (16PF) are the ones which have been most intensively investigated by Cattell and his associates. These factors are considered to represent 'source traits' which are the most important units in Cattell's theory. They are the basis of the enduring regularities observed in behaviour, and it is through the interaction of these source traits that the more readily observed 'surface traits' of an individual are determined. These surface traits correspond more closely to the ordinary personality descriptions which the layman uses. Cattell's preference for oblique factor analysis results in small correlations between the primary factors and if a further factor analysis is carried out on the primary factors then a number of second-order factors can be extracted.

The two most important second-order factors are given the

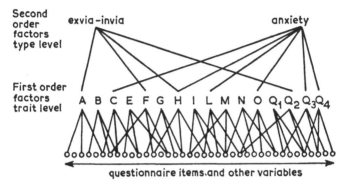

Fig. 4.1 *The hierarchical organization of the factors derived from questionnaires* (after Cattell, 1965, p. 118)

names exvia-invia and anxiety (cf. Eysenck's extraversion-introversion and neuroticism). The second-order factors represent broad reaction tendencies which, in Cattell's view, only influence actual behaviour through the intermediate primary factors which are considered to be more accurate than the second-order factors in describing and predicting behaviour. A diagrammatic representation of the hierarchical organization of Cattell's theory is contained in Fig. 4.1.

Although he believes that general personality factors and abilities remain fairly stable over a long period, Cattell recognizes that motivational variables are likely to fluctuate from moment to moment, particularly under the influence of situational variables. The two main factors which bring some regularity to this situational variability are 'states' and 'roles' of the individual. The more enduring trait descriptions of an individual's personality may have to be considerably modified when his behaviour is influenced by a temporary state such as fatigue, happiness, fear or intoxication. A man who is tired, frightened or drunk may seem to 'act out of character'; in other words, predictions of his behaviour on the basis of trait factors alone are likely to be misleading. Cattell regards personality variables as one of a set of conditions which influence behaviour: 'Personality is what determines behaviour in a defined situation and a defined mood' (Cattell, 1965, p. 27).

In the description of Cattell's basic theory we have not so far attempted to explain how individual differences in personality arise. It is generally accepted that environmental and hereditary influences determine behaviour and that their effects are inextricably interwoven. However, the relative contributions of heredity and environment will differ with regard to different types of behaviour. Cattell and his associates have developed a special research technique, called 'Multiple Abstract Variance Analysis' in order to investigate the relative contributions of heredity and environment to different traits. The amount of variability contributed by environmental or hereditary influences varies considerably for different traits, for example, intelligence has been estimated to be largely determined by heredity (80%) by Cattell, whereas neuroticism or Ego Weakness has been estimated to be less influenced by hereditary factors (30–40%). However, Cattell recognizes that environmental and hereditary influences interact, in each factor, to a greater or lesser extent.

Assessment

A number of questionnaires have been developed by Cattell and his associates but the most widely known is the 16PF Questionnaire (Cattell, R. B., Eber, H. W. and Tatsuoka, M. H., 1970). This test has five alternative forms. The most widely used forms, A and B, contain 187 items. Two examples are set out below.

I I like to watch team games
    a) yes   b) occasionally   c) no.
2 I prefer people who
    a) are reserved   b) (are) in between
    c) make friends quickly.

The subjects' responses are scored so that a score from one to ten is obtained on each of the sixteen trait dimensions shown in Figure 4.2 below.

Data on test-retest reliability (see p. 19) are not presented, because Cattell assumes that normal variations in traits occur

Low Score Description	Factor	High Score Description
Reserved, detached, critical	A	Outgoing, warmhearted
Less intelligent, concrete thinking	B	More intelligent, abstract thinking
Affected by feelings, easily upset	C	Emotionally stable, faces reality
Humble, mild, accommodating	E	Assertive, aggressive, stubborn
Sober, prudent, serious	F	Happy-go-lucky, impulsive, lively
Expedient, disregards rules	G	Conscientious, persevering
Shy, restrained, timid	H	Venturesome, socially bold
Tough-minded, self-reliant	I	Tender-minded, clinging
Trusting, adaptable	L	Suspicious, self-opinionated
Practical, careful	M	Imaginative
Forthright, natural	N	Shrewd, calculating
Self-assured, confident	O	Apprehensive, self-reproaching
Conservative	Q^1	Experimenting, liberal
Group-dependent	Q^2	Self-sufficient
Undisciplined self-conflict	Q^3	Controlled, socially precise
Relaxed, tranquil	Q^4	Tense, frustrated

Fig. 4.2 *From the 16PF Questionnaire*
(© 1956–1967 by the Institute for Personality and Ability Testing, Champaign, Ill., USA)

over time and low test-retest reliabilities are, therefore, to be expected. The validity of the 16PF factors has been seriously questioned in recent years, particularly since developments in computer technology have made it possible to carry out large-scale factor analyses of 16PF Questionnaire data. This is a crucial issue because the 16PF is so closely bound-up with

63

Cattell's theory of personality that a demonstration that 16PF factors are not replicable could seriously damage the basic theory.

A number of investigations of the 16PF test have found factors which do not appear to match those of Cattell. A study by Eysenck and Eysenck (1969) which involved the factor analysis of 16PF items found a smaller number of factors. Howarth and Browne (1971) factor analysed the 16PF test data of over 500 students and they concluded that no clear factor structure emerged. The factor loadings seemed to bear no relation to the scales of the 16PF and they were critical of the test's validity. The match between the 16PF items and the constructs the test purports to measure is not strong and a number of studies do not fully support the framework upon which the 16PF is based. The results are more encouraging for the second-order factors of Exvia and Anxiety and it has been suggested that the primary factors may be of lesser importance.

It should be pointed out that all the studies so far undertaken appear to have used an earlier version of the 16PF Questionnaire. There is an improved 1967 version in which a third of the items have been changed, and Cattell (1974) considers that the above criticisms do not necessarily apply to this new version. No doubt we can expect more studies of this type in the next few years, which may resolve some of the unanswered questions about the validity of the 16PF factors. For the moment its applicability with both individuals and groups remains questionable.

Applications
Abnormal psychology. Unlike many other personality theories Cattell's has not been strongly linked to the area of abnormal psychology. He views neurotic problems largely in terms of an individual holding an extreme position on a trait and being exposed to unfortunate life experiences and no dichotomy between abnormal and normal personality is assumed. Cattell believes that such a dichotomy may be shown to exist for certain forms of mental disorder such as schizophrenia. Most

research in the clinical field using Cattell's theoretical concepts has been concerned with establishing typical 'personality patterns' for various diagnostic groups with the 16PF Questionnaire. The most consistent finding for all abnormal groups is a low score on Factor C (Ego strength or emotional stability) of the 16PF test.

In his theory of maladjustment Cattell attaches considerable importance to the role of conflict. An individual may experience conflict when a drive is stimulated and then blocked; this may occur when a behaviour satisfies one drive while leaving another frustrated (Approach-Avoidance Conflicts). In attempting to counsel an individual in a state of conflict Cattell would advocate that the nature of the conflict should be mapped out by means of questionnaires and objective tests. The counsellor is then in a position to advise his client on the roots of his tension and to suggest ways of alleviating his symptoms. Psychotherapeutic approaches which seem to hold out the promise of unlimited change are criticized by Cattell because he considers that trait levels are in part constitutionally determined and could impose limits on what may be achieved. This approach to counselling does not seem to be widely followed in clinical practice and a study by Williams *et al.* (1972), which examined the utility of the 16PF in the clinical situation, concluded that it is not an adequate diagnostic tool in a mental hospital assessment programme. This is partly because of its susceptibility to distortion and also because it was unable to provide adequate measures of more extreme forms of abnormal behaviour.

Education. In education Cattell's influence has probably been greater. The description of the school psychologist as someone who 'applies learning theory to the curriculum, psychometrics to the improvement of examinations and personality theory to the treatment of maladjustment and the problem of personality and character education' (Cattell, 1965, p. 291) would probably receive wide support, though Cattell himself recognizes that this 'ideal' state of affairs may seldom occur in practice.

65

It has been strongly argued by Cattell that intelligence and achievement measures must take account of personality factors if they are to be of value. This rarely happens at present. In devising 'Culture-Fair' tests of intelligence Cattell hoped to establish the existence of a component of intelligence which was distinct from the acquired ability derived from the educational process. The Culture-Fair tests were thought to measure 'Fluid' intelligence which was largely the product of heredity, in contrast to 'Crystallized' intelligence which was a complex of skills arrived at by the interaction of fluid intelligence with education (Cattell, 1963). Support for this theory has come from Cattell's factor analytic studies of intelligence, which indicate that when personality variables are included in investigations the two predicted factors emerge.

Evaluation
The adequacy of Cattell's theory of personality depends on the ability of factor-analysis to uncover basic personality dimensions. This ability has been seriously questioned by a number of psychologists. In his review of the applications of factor analysis to personality research, Lykken (1971) suggests that there are three main uses of factor analytic techniques. The first two concern data reduction and hypothesis testing and these are considered to be appropriate, but rarely used, applications of the method. The third use is that of Cattell and his associates which attempts to lay bare the assumed hierarchical structure of personality and to unravel its basic dimensions.

If Cattell's theory is to gain wide acceptance among psychologists, it must be possible to demonstrate that there is a fairly stable number of primary factors which emerge from different forms of measurement (chiefly questionnaires and objective tests) with different subjects. In addition, it should be possible to name the factors unambiguously and to demonstrate that they have some psychological as well as mathematical reality.

There is unfortunately no clear agreement among personality researchers about the number of factors needed to account for the data obtained by Cattell. This is perhaps not surprising when we remember the importance of intuitive factors in the process of factor analysis. Thus, as Lykken (1971) points out, many of the disagreements are reduced to questions of personal opinion. It is difficult to avoid this pitfall in the area of personality research but Lykken argues that tests on 'artificial data' might demonstrate whether or not the technique of factor analysis really does aid our understanding of complex material. In one test he factor analysed road-test data on a hundred cars and found four factors which accounted for 86% of the common variance (see p. 111). It proved possible to attach tentative names to two of the factors such as a 'performance' factor, and an 'economy' factor though the others were more difficult to interpret. The important point is whether this data gives someone who is relatively ignorant about what goes on 'under the bonnet' any useful information about the structure of a car which he did not have already. Lykken concludes that the answer is obviously 'No'.

The major criticisms of Cattell's theory are that the assumptions of the factor analytic model force us to oversimplify personality. In addition the evidence of stability and replicability of traits between different subjects and across situations is unsatisfactory. Brody (1972, p. 22) has carefully reviewed the available evidence and concluded that 'the use of factor analytically derived traits based on first order factors as a descriptive system for personality rests on faith rather than facts.' Fortunately, the position with regard to second-order factors seems to be more encouraging. A number of studies have found that the Exvia-Invia and Anxiety second-order factors are relatively stable.

Eysenck's type theory

The personality organization described by Eysenck has many similarities to that of Cattell. The chief difference concerns the level of organization which is considered to be most important, with Eysenck stressing the type or second-order factor level as opposed to the trait level. (Fig. 4.3).

There are other major differences in the two approaches. Eysenck, while sharing Cattell's view that biological factors are involved in determining personality, has gone much further in providing causal explanations. Eysenck's theory has generated such a large volume of experimental research that it will only be possible to consider a few selected areas in the

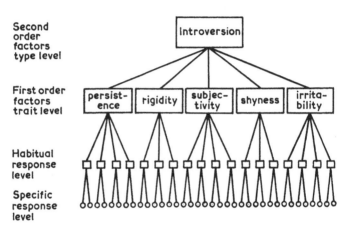

Fig. 4.3 *Eysenck's model of personality structure in relation to the Introversion dimension* (after Eysenck, 1953, p. 13)

course of this chapter. The areas chosen are those which bear most directly on the central aspects of the theory.

Unlike Cattell much of Eysenck's early work grew out of his interest in abnormal psychology and psychiatry. He acknowledges a debt to the writings of Galen, Kant, Wundt and Jung all of whom shared the view that individual temperaments could be described by a small number of different types. The basic dimensions are shown in Figure 4.4 and the inner circle represents the four classical temperaments described by Galen almost 2,000 years ago.

Eysenck has supplemented his factor analytic work with the method of criterion analysis. The use of criterion analysis is intended to reduce the arbitrary nature of the process of naming factors. Having selected a dimension such as Extraversion-Introversion, along which individuals are assumed to vary, the next step is to collect test data from groups of individuals who are expected to differ widely on the dimension. Thus if certain diagnostic groups of psych-

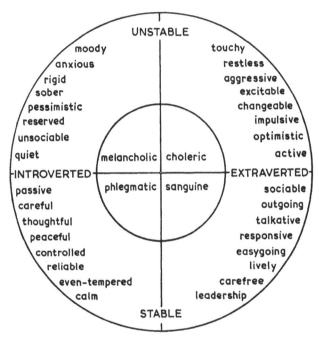

Fig. 4.4 From Eysenck, 1965, p. 54

iatric patients are presumed to be highly extraverted while others are considered to be highly introverted, then analysis of the test data should place them at different ends of the Extraversion-Introversion dimension.

In his original study of seven hundred neurotic soldiers Eysenck (1947) found that factor analysis of thirty-nine items of personal data, including personality ratings, resulted in the establishment of the two basic dimensions of Extraversion-Introversion and Neuroticism-Stability. These dimensions were supported by further research, with literally thousands of subjects. In a later investigation with psychiatric patients Eysenck (1952) established a third dimension, unrelated to Extraversion and Neuroticism, which he labelled 'Psychoticism'. Thus Eysenck's personality system is made up of four dimensions, viz:

Extraversion	–	Introversion
Neuroticism	–	Stability
Psychoticism	–	Stability
Intelligence		

Although intelligence is the only cognitive dimension in this structure it is not regarded as a pure dimension and the possibility of there being other important cognitive dimensions is not denied. Until recently the Psychoticism dimension received relatively little attention and most of Eysenck's research and theory construction has been concerned with the Extraversion and Neuroticism dimensions.

Eysenck uses the terms extravert and introvert in ways which resemble but are much wider than their use in everyday language.

The typical extravert is sociable, likes parties, has many friends, needs to have people to talk to, and does not like reading or studying by himself. He craves excitement, takes chances, often sticks his neck out, acts on the spur of the moment, and is generally an impulsive individual. He is fond of practical jokes, always has a ready answer and generally likes change; he is carefree, optimistic, and likes to laugh and be merry. He prefers to keep moving and doing things, tends to be aggressive, and loses his temper quickly. Altogether, his feelings are not kept under tight control, and he is not always a reliable person.

The typical introvert, is a quiet, retiring sort of person, introspective, fond of books rather than people; he is reserved and distant except with intimate friends. He tends to plan ahead, 'looks before he leaps', and distrusts any impulse of the moment. He does not like excitement, takes matters of everyday life with proper seriousness, and likes a well-ordered mode of life. He keeps his feelings under close control, seldom behaves in an aggressive manner, and does not lose his temper easily. He is reliable, somewhat pessimistic, and places great value on ethical standards. (Eysenck, 1965, p. 59)

These descriptions relate to the extreme ends of the dimension and very few individuals would fit them completely. Most people fall somewhere in between and are neither strongly extraverted nor strongly introverted. The Extraversion dimension may be subdivided into two main components; sociability and impulsiveness, though most research has concentrated on the higher-level concept of Extraversion.

The Neuroticism dimension is similar to the notion of emotional instability. Those individuals who fall at the extreme neuroticism end of the dimension tend to be more prone to worries and anxieties and more easily upset. They are also likely to complain of headaches, and sleeping or eating difficulties. Although they may be more likely to develop neurotic disorders under stressful conditions the frequency of such problems is low and most individuals function adequately in their work and in their family and social life.

The causal factors in Eysenck's theory are firmly rooted in biology and are highly complex (Eysenck, 1967). Personality factors are not thought to be inherited directly but rather an individual inherits a particular type of nervous system which predisposes him to develop in a particular direction. The final shape of the personality will be determined by the interaction between an individual's biological predisposition and the environmental influences that he encounters during his life. The original form of Eysenck's theory relies heavily on the work of Pavlov (1927) and Hull (1943). The ease and stability with which an individual forms conditioned responses (see A3) is considered to be related to the balance between excitation (activating) and inhibition (dampening down) processes within the central nervous system. The learning of stimulus-response connexions is favoured by a strong and rapid build-up of excitation in the nervous system and a tendency for inhibition to develop slowly and weakly. Introverts are considered to have inherited such a nervous system and are, therefore, capable of strong and rapid conditioning. Extraverts, on the other hand, are considered to form conditioned responses slowly and weakly.

The physiological basis for this difference in conditionability has been related to the functions of the Reticular Activating System (RAS) in the brain (see A2). The main function of the RAS appears to be to maintain the individual in an optimum state of 'arousal' or alertness. In addition to the activation role of the RAS it also has an inhibitory or dampening function. Claridge (1967) could find no simple relationship between levels of Extraversion and Neuroticism and physiological arousal: however, there did appear to be a complex interrelationship between arousal and an individual's position on the two dimensions. Thus introverted neurotics appear to be more highly aroused than less introverted and less neurotic patients.

Emotionality or neuroticism is related to the reactivity of the autonomic nervous system (see A2). Individuals with more labile autonomic nervous systems are liable to respond strongly to unpleasant or frightening experiences by increases in heart-rate, muscle tension, sweat-gland activity and so on. Individuals high on neuroticism will tend to have low thresholds of emotional arousal. This will lead to the more frequent activation of their autonomic nervous systems which in turn will trigger the RAS. Thus the RAS will be more often in a state of arousal for individuals with high scores on neuroticism. This means that such individuals will tend to resemble introverts who are generally more 'aroused' than extraverts.

The biological aspects of the theory attempt to lay the basis for the assumed differences in conditionability of Introverts and Extraverts. However, not all studies of conditioning have found the expected differences between Introverts and Extra-

72

verts, viz. that Introverts condition more easily and stably than Extraverts. There are a number of important inconsistencies in the research findings and it has not always proved possible to replicate experiments. In addition, the theory cannot accommodate some of the established facts about conditioning without making *ad hoc* assumptions. Brody (1972) concludes that 'it might be reasonable to suggest that the relationship between introversion and extraversion and conditioning is not clearly understood.' (p. 61)

This lack of clarity inevitably raises fundamental doubts about Eysenck's subsequent theorizing on the socialization process and the development of neurosis, since it rests very largely on the assumed differences in conditionability of Introverts and Extraverts.

Assessment

Eysenck's attempts to devise a reliable questionnaire for measuring the dimensions of Extraversion (E) and Neuroticism (N) originated from his observation that some earlier scales could discriminate between two clinically neurotic groups which resembled his description of introverts and extraverts. He then drew up a questionnaire of 261 items which contained items from earlier scales and factor analysed the responses of 200 male and 200 female normal subjects. This enabled him to reduce the number of items to 48, which make up the Maudsley Personality Inventory (MPI). Although the MPI has been used extensively in personality research it has now been largely superseded by the Eysenck Personality Inventory (EPI) (Eysenck and Eysenck, 1964).

The EPI consists of two parallel forms (A and B) which contain 57 items scored YES or NO, e.g. Do you often long for excitement? Yes ——; No ——. This test is intended to replace the MPI though the content of the items is very similar. A strongly worded critique of the EPI and of forced choice questionnaires in general is contained in Heim (1970). Her main conclusions are that a few simple, highly intimate questions to which the subject is forced to respond YES or NO can hardly be expected to do justice to the complexities of personality. The Lie scale items which are included to screen out subjects making 'socially desirable responses' are criticized for their lack of subtlety, e.g. 'Are all your habits good and desirable ones?' Yes ——; No ——. Rump and Court (1971) found that neuroticism items can be contaminated by social desirability effects, even when there is no obvious advantage in 'faking good', such as when it is used with normal subjects for research purposes. The effect is also noted when the scale is used for clinical assessment and the use of the EPI as a screening test with individuals may, therefore, be suspect.

The validation of the EPI has been largely in terms of ratings of known groups (i.e. individuals at the extremes of the E and N dimensions) (Eysenck and Eysenck, 1963). This leaves some doubt about the status of the majority who occupy the middle ground. Gibson (1971) attempted to overcome this difficulty by asking students to complete the EPI as a rating assessment on an unselected group of their friends and he found significant overall correlations between self-ratings and ratings by others on the E and N scales. This finding offers some support for the validity of the EPI when used with unselected subjects who may be located at any point on the two dimensions. An as yet unpublished revision of the EPI contains a total of ninety-five items measuring P (Psychoticism), E (Extraversion), N (Neuroticism), and L (Lie Scale). The status of the Lie Scale is yet to be clearly established, but Eysenck *et al.* (1974) have suggested that it may represent a 'stable personality trait' which is independent of P, E and N.

Applications

Eysenck has attempted to apply his theory to a very wide range of human behaviour, including politics, accidents, sexual behaviour, and the effects of drugs on behaviour. The two areas which have generated considerable interest and controversy have been chosen for more detailed discussion, viz. applications to psychiatric diagnosis and criminal behaviour.

Psychiatric Diagnosis. The present state of psychiatric diagnosis, assessment and treatment has been strongly condemned:

'We thus arrive at a rather dismal picture in which unreliable, invalid and largely arbitrary tests are used to arrive at unreliable, invalid and largely arbitrary diagnoses which are irrelevant to the methods of treatment to be used, methods which in turn are quite ineffective.' (Eysenck, 1970, p. 171)

The attack on the present categorial, classificatory approach to psychiatric disorders has come from a number of different quarters, but Eysenck's assault has been perhaps the most sustained. It is also backed by an impressive amount of experimental work in the fields of personality assessment and learning with normal subjects. Eysenck argues for a dimensional approach to psychiatric diagnosis with two independent dimensions of neuroticism and psychoticism each forming a continuum from extreme abnormality to normality. The evidence for Neuroticism as a normal personality dimension is well established but Psychoticism has been neglected until recently. Although Claridge and Chappa (1973) found evidence supporting Psychoticism as a normal personality dimension they also found a number of similarities between low N subjects and high P subjects on several psychological and physiological measures. They suggest that for some individuals low neuroticism scores may reflect a kind of emotional blunting associated with certain forms of psychotic personality not measured by Eysenck's P scale.

Within the broad group of neurotic disorders Eysenck distinguishes between dysthymic disorders (depressives, obsessionals and phobics), who are all characterized by high Neuroticism and high Introversion, and hysterics and psychopaths who are considered to be high on Neuroticism and Extraversion (see F3). A number of studies have confirmed these predictions for dysthymics but the position of hysterics seems to be rather different from that originally suggested by Eysenck. Hysterics appear to be somewhat lower than dysthymics on the Neuroticism dimension and normal on Extraversion, i.e. neither highly introverted nor highly extraverted (McGuire *et al.*, 1963).

The dysthymic disorders are considered to be due to the formation of conditioned emotional responses (learned anxiety and fear reactions) to normally neutral stimuli and these conditioned responses and the motor movements made in response to them constitute the neurotic disorder. Hysterics, because of their relative extraversion are assumed to condition less easily, except for situations where very strong stimuli are present. In support of this view Eysenck argues that hysterical reactions commonly occur in situations of extreme stress, e.g. on the battlefield. Psychoticism is also regarded as a continuous dimension ranging from normality to extreme abnormality and Eysenck believes that the traditional psychiatric distinction between schizophrenic and manic-depressive psychoses has not been well supported by the evidence, though the situation remains unclear. So far Eysenck has made little attempt to relate Psychoticism to a biological basis.

Although the theory has been extensively linked with behaviour therapy approaches to the treatment of neurosis (see Ch. 7) its impact upon therapeutic practice has been quite small. It is only possible to quote anecdotal evidence (Eysenck, 1970) of direct applications to treatment and most behaviour therapists do not appear to utilize the concepts of Extraversion and Neuroticism in their work. A number of suggestions derived from Eysenck's theory have been made in relation to such questions as the use of long or short treatment sessions but as yet few have been experimentally investigated in the clinical situation. Equally important might be the role of levels of E and N in forecasting the outcome of various behaviour therapy procedures. The lack of evidence in this area has resulted in a general disillusionment of behaviour therapists with standardized personality tests as predictors of response to treatment or measures of change.

Criminal behaviour. The theory of criminal behaviour proposed by Eysenck (1964) relies heavily on the assumed differences in conditionability between Extraverts and Introverts. In view of their weaker conditionability Extraverts are under-socialized and have less well developed 'consciences'. In addition high Neuroticism and more recently high Psychoticism (Eysenck and Eysenck, 1970) levels have also been linked to criminality, though the rationale for this is less clear. It is individuals high on both Extraversion and Neuroticism who are likely to show criminal behaviour. Cochrane (1974) has reviewed the available studies of Eysenck's theory by questionnaire measures on prisoners and control groups and concluded that although prisoners are generally found to be higher on Neuroticism they are not found to be higher on Extraversion. Indeed a number of studies have shown criminals to be more introverted than controls. Eysenck (1974) attempts to explain these uncomfortable findings by suggesting that the questionnaires used (EPI and PEN) have been largely measuring the 'sociability' component of Extraversion whereas it is the 'impulsivity' component which is related to conditionability. Unfortunately the evidence to support this statement with regard to conditioning is rather sparse. Cochrane argues that the theory has been discredited in its original form and this raises serious doubts about the adequacy of Eysenck's general theory of 'socialization'.

Evaluation
It was originally hoped that a factor analytic theory which concentrated on a small number of second-order 'type' factors would prove more adequate than Cattell's trait theory but there have unfortunately been a number of disappointments. The assumed unitary nature of the two dimensions has been found incapable of accommodating some of the experimental evidence originally derived from the theory. Eysenck has become increasingly convinced of the importance of a third dimension, Psychoticism, in normal personality and there are hints at a fourth dimension. However, the dimensions have not always shown the independence that was originally claimed for them.

The theory is at its weakest in relation to the socialization process. In the absence of unequivocal evidence that Introverts will condition more rapidly than Extraverts, there is little justification for making broad generalizations about the development of differences in social behaviour for the two groups on the basis of differences in conditionability. This also puts at risk Eysenck's attempt to link social behaviour to 'causal' biological factors. Also Eysenck over-emphasizes the role of classical conditioning in socialization and neglects the importance of social learning variables (see Ch. 7).

In the present state of personality research it is natural that a theory which is presented as explicitly as Eysenck's will frequently have to be revised in the light of new evidence. The theory is still useful in generating testable predictions, though it has found some of the recent experimental evidence rather indigestable. Eysenck's personality theory provides a useful model for the general differences found among groups of individuals. Greater attention to other principles, particularly social learning, and cognitive factors, other than intelligence, will be required before any theory of this type can be expected to have wider application to the understanding of individual behaviour.

At present there seems to be little justification for Eysenck's suggestions that psychiatrists, educationalists, or parole boards should base their recommendations for treatment, teaching, or probation on measures of Extraversion, Neuroticism or Psychoticism. The questionnaires which Eysenck has developed may continue to be useful tools for personality investigators but their practical value in relation to groups or individuals remains to be demonstrated. Despite these reservations, the main body of Eysenck's theory seems certain to prove a fruitful source of ideas for investigation in the field of personality for many years to come.

5
Narrow band
theories (I)

In Chapter 1, a distinction was made between comprehensive and narrow band theories, the latter being concerned with more specific and restricted content area. There is no rigid distinction between them, the difference being mainly in degree rather than in kind. The four theories discussed here were chosen because they seem to be representative of many narrow band theories. This chapter will examine two cognitive theories; the following chapter will examine two motivational theories, followed by a general critical look at trait approaches to personality.

(1) *Internal-external locus of control*

An early exponent of the 'social learning' or behavioural approach to personality (Ch. 7) was J. B. Rotter (1954), but his work is now best known not for his behavioural approach, but for an offshoot of his earlier work, the dimension of internal versus external control of reinforcement. Apparently he never regarded this as a personality dimension, but this is exactly how it has been used by other psychologists.

Rotter maintains that people can be classified along the following continuum: at the Internal end are people who view their behaviour and what happens to them as directly under their personal control. At the other are the Externals, who view their behaviour as influenced by events that are completely out of their control and due to such factors as chance, powerful figures in authority, and fate. The internal will be confident that he can bring about changes in his environment and in his own behaviour, whereas the external will feel comparatively powerless to produce change.

The consequences of a person's behaviour establish an expectancy that the same consequences will result from similar behaviour in the future; if different consequences occur, the expectancy will be reduced, but if similar consequences consistently follow, the expectancy will be strengthened. By such means, a child learns to distinguish behaviours that bring about predictable results from behaviours that do not. He will gradually develop a relatively stable attitude or series of expectancies regarding the locus of control of his own behaviour (that is, whether the results are due to his own actions, or to some other factor).

Individuals, therefore, develop expectancies concerning the consequences of a wide range of learning situations, and because people's learning experiences differ, the same situation may be perceived differently by different people. Eventually expectancies become generalized, so that people generally feel in control or otherwise across many different situations.

Expectancies can arise from instructions as well as from direct learning. It has been shown that if subjects are told that the results in a gambling situation are largely due to chance, and others are told the results are due to skill, very different expectancies and outcomes result, even though the 'pay-off' for each group is the same. If people attribute outcome to skill, their bets increase with successes, and fall with failures. But if they attribute outcome to chance, the opposite pattern tends to emerge – bets are increased after failure and decreased after success. It appears that subjects apply the mythical 'law of averages' to chance situations and reason that after several failures, the chances of success increase, and that after several successes the chances of failure increase. This is well known in gambling circles as the 'Monte Carlo Fallacy'.

Because of their belief that they are in control of their own destiny, internals could be expected to be more independent, effective, achieving and dominant, and there is much experimental evidence to support this. Externals, however, feeling that they have little control over their environment and, therefore, being unable to take effective remedial action, would be more likely to feel high levels of anxiety and report more neurotic symptoms. Again there is some experimental support for this. It also appears that externals are more suspicious and dogmatic than internals. Similarly, one might expect that the poor and the downtrodden would feel comparatively powerless to change their environment, and would, therefore, have higher external scores. Many studies have shown that American negroes, the unemployed, the poor and other minority groups, tend to have higher external scores. People whose life is restricted, with little power and few opportunities to change their lot, tend to develop an external attitude to life. This can, of course, be to the individual's advantage; an internal negro is likely to accept self-blame for his underprivileged condition, whereas an external negro would more likely (and adaptively) attribute his condition to society in general.

Externals are lacking in motivation to achieve difficult goals and have few aspirations, since their efforts are seen as bearing little or no relationship to outcome. 'Purposelessness' and a 'meaningless' existence are also said to characterize the behaviour of people who are clinically depressed, and it has, therefore, been put forward that there should be a relationship between externality and depression. Interestingly, Seligman (1973) has proposed a similar analysis of depression in terms of 'learned helplessness', based on a series of experiments in which dogs were given electric shocks which they were powerless to avoid. The dogs developed behavioural and bodily symptoms strikingly similar to those found in human depression. Abramowitz (1969) has examined this relationship

81

in a group of university students, and found a significant correlation ($+ \cdot 28$) between externality and depression. Other studies relating externality and proneness to depression have produced somewhat equivocal results. The nature of this relationship requires considerably more elucidation before the findings could be applied.

Finally, Harris and Nathan (1973) found that parents who rated their children's problems as arising from sources outside their control (e.g. 'Mary is temperamentally insensitive', or 'John is brain-damaged') were significantly more external than parents who accepted personal responsibility for the problems.

To conclude, internal-external orientation reflects how far a person expects that his actions will be instrumental in bringing about change. The internal believes that he has personal control, whereas the external attributes his experiences to fate, luck, or factors beyond his control. Orientation along this dimension results from learning experiences. The internal is more likely to have high aspirations and persistence, to participate in political activity, to be resistant to persuasion, to exert influence over others, and to be independent. The external tends to have the opposite characteristics, refusing to accept personal blame, and being prone to low moods.

Assessment

Several ways of measuring internal-external orientation have been developed, and the most intensively investigated is that of Rotter (1966). This questionnaire (called the I-E scale) has twenty-nine items, each item consisting of two statements, and the subject is required to select the statement that most nearly describes what he believes to be the case. Some examples are:

2 a) Many of the unhappy things in people's lives are
 — partly due to bad luck.
 b) People's misfortunes result from the mistakes they make.

28 a) What happens to me is my own doing.
   b) Sometimes I feel that I don't have enough control
   — over the direction my life is taking.

In the above examples, the underlined items would count towards the external score.

The overall reliability of the I-E scale has been found to vary between 0·49 and 0·85, normally towards the higher end of this range. The validity has recently been questioned however; it has been suggested that it is socially desirable to be seen as possessing many of the characteristics of the internal and that some people may 'fake good' on the test in order to present themselves in a favourable light. Joe (1971) summarized recent work suggesting the existence of a relationship between I-E score and social desirability. It is possible, therefore, that, at least in part, the I-E scale is a measure of social desirability, rather than I-E orientation. Joe (1971) also suggests that I-E score is correlated with such non-personality variables as age, sex, intelligence and political views. Thus, the purity of the I-E scale as a measure of internal-external orientation has been seriously questioned.

The I-E scale has been factor analysed several times, and there is general consensus that it does not measure a unitary factor, different studies having produced different factors. For example, Gurin *et al.* (1969) found four factors, which they called Control Ideology (how much control most people in society have), Personal Control (how much control one has personally), System Modifiability (how far society's problems can be overcome) and Race Ideology (containing most of the items that refer to race). Subsequent studies factor analysing the I-E scale have not found the same factor structure, but there is some general agreement about the existence of the factors of Personal Control and System Modifiability. Collins (1974) has argued that an extremely external person could be a person who feels he lives in a complex, difficult world, or who thinks that effort and ability are unrewarded, or who thinks that life is governed by fate, or that governments are unresponsive. Any of these differing beliefs would result in

83

high external scores, and, therefore, externals are not necessarily a homogenous group.

There is then some doubt about precisely what is measured by the I-E scale. Accordingly, to use the total I-E score in predicting other behaviour may not be appropriate.

In sum, the reliability of the I-E scale has generally been found to be satisfactory. However, some doubts have been raised concerning its validity, particularly in respect of its relationship to social desirability, and its factorial 'purity'.

Applications

Psychiatric diagnosis. Harrow and Ferrante (1969), in a very wide-ranging study, administered the I-E scale to groups of psychiatric in-patients from a number of different diagnostic categories. The average for the patient group as a whole did not differ significantly from the normal range of scores, but scores of the schizophrenic and the manic subjects were significantly more external than the rest. Harrow and Ferrante also retested some of the patients six weeks later, and found a slight but nonsignificant trend for the patients in general to become more internal. Trends again differed amongst the diagnostic groups, the schizophrenics and manics becoming more external, the depressives and others becoming more internal. This was consistent with some previous studies, and with their expectations that in most groups a move towards psychological well-being would be accompanied by increasing internality. However, the differences found tended to be of small magnitude and certainly too trivial for the I-E scale to be recommended as an instrument to monitor improvement in psychiatric patients, or for use as an aid to psychiatric diagnosis.

As regards general psychological adjustment, it is commonly found that externals tend to be more anxious, less able to deal with frustration and less concerned with achievement. Thus the external is less 'psychologically healthy', but it has also been suggested that both extreme externals and extreme internals are more maladjusted than individuals in the middle range (Rotter, 1966).

84

Suicide Proneness. Characteristics of a typical 'suicide prone' individual are very similar to those of the externally orientated individual. This similarity has led Williams and Nickels (1969) to investigate the relationship between externality and suicide proneness (as measured by rating scales and questionnaires) and they reported correlations of the order of $+\cdot20$; unfortunately, this low correlation would contribute little to the detection of the suicide-prone individual. Furthermore, the I-E scale does not appear to have been related to actual suicidal behaviour, but to psychometric assessments of it. A more direct study of the relationship may prove to be more fruitful, but more recent studies (e.g. Lambley and Silbowitz, 1973) have suggested that the I-E scale cannot easily differentiate between those who have contemplated suicide and those who have not; accordingly the scale's usefulness in detecting suicide proneness (and thereby hopefully preventing its occurrence) must remain in doubt.

Evaluation
The dimension of internal-external locus of control has generated an enormous amount of research over the last decade. Theoretically its importance lies in the suggestion that expectancies can play a major role in determining the effects of reinforcement. There are, however, a number of difficulties associated with this concept. Some major ones have already been referred to: the association with social desirability, the weak predictive power resulting from low correlations, and finally the factorial 'impurity' of the I-E scale. A further problem is that results tend to be very inconsistent; some studies show internality to be related to a particular variable, other studies show externality to be related to the same variable, and still others show no relationship at all.

Many of these problems may result from the use of the total I-E score in research investigations. Although there are some important differences between them, many of the factor analytic studies point to the existence of two main locus of control factors – Personal Control and System Modifiability.

Separate examination of the effects of these two largely independent factors may well sharpen up the predictive utility of the I-E scale, but fuller research is required to demonstrate this.

(2) *Field dependency–independency*

In the recent history of psychology, there have been many attempts to approach the study of human behaviour via the processes of perception. Perception is more than a passive receiving of incoming stimulation from our eyes, ears and skin senses. It is a very active process of selecting which stimuli will be admitted, and interpreting these stimuli according to hereditary mechanisms, previous experience and the motivational state of the person (see A4).

It is this active aspect of perception that has led many psychologists to postulate that if standard stimuli are presented to a series of subjects, the different ways in which the stimuli are seen and reported will provide useful cues in the assessment of the subjects' personality. If the stimuli are ambiguous or unusual, this would give the subject more freedom to respond and should provide a more sensitive and wide-ranging measure.

The work of Witkin and his colleagues on field dependence–independence is to the fore in the modern theories using this approach. Witkin's theory arose from some early experiments in the 1940s concerned with judging whether a straight rod was seen to be vertical when set in various planes. Subjects who were most accurate in perceiving verticality in the rod experiments also tended to be most accurate in similar tasks. The most salient features that the tasks had in common were that they involved assessment of the ability to perceive an object in relation to and separate from its surroundings (see pp. 88–9 for details of these tasks). For some people perception of the object was strongly influenced by the background or surrounding field, whereas others were not so influenced and could differentiate more easily between a part and its larger, surrounding field. The former have been called 'field-dependent'; the latter, 'field-independent'.

Witkin then suggested that such differences in perception may be consistent characteristics of the individual in a variety of other perceptual tasks. The most extensively investigated of these is the embedded figures test, which requires the subject to recognize standard figures which are 'hidden' in a complex background. Again it was found that people differed systematically and consistently in their ability to escape the contextual constraints of the backgrounds. Subsequent work has indicated that such 'styles' are not confined simply to perceptual processes, but also exert an influence in intellectual and social activities. This aspect has recently attracted considerable attention, and Witkin's theory of 'differentiation' is now regarded as a cognitive rather than a perceptual theory of personality. (It should be noted that others (e.g. Gardner *et al.*, 1959) have also produced a theory of differentiation, which is similar to that of Witkin, but differs in some important respects.)

In a variety of activities and settings, Witkin and others have demonstrated that people have different 'cognitive styles'. These styles may be represented as positions on a continuum with, at one end, people with a consistent tendency to interpret events in an undifferentiated gross manner, for whom the organization of the field exerts a powerful influence over the interpretation of its constituent parts. At the other end of the continuum are people with a consistent tendency to interpret events in a detailed, organized way for whom it is a straightforward matter to regard quite independently constituent parts of the whole field. These extreme versions of cognitive style have been called 'global' and 'articulated' respectively. Differences in differentiation are assumed to reflect basic properties of the individual's personality and should, therefore, manifest themselves in a variety of situations.

It may help to reinforce the understanding of the varied terms used in Witkin's theory by stressing that, on the whole, the terms field-dependent, global and undifferentiated are largely synonymous, as are the converse terms field-independent, articulated and differentiated.

A relatively well differentiated person would have a clear idea of his own beliefs, needs, characteristics and so on, which would not only be seen as separate from each other, but also as very clearly distinguishable from those of other people. With the person of undifferentiated style, internal dynamics would overlap and influence each other, and it would be difficult to keep them entirely separate from those of other people. Accordingly, the global, undifferentiated, field-dependent person would rely on others for the formation and maintenance of attitudes, and would be much more susceptible to social influence and less able to rely on his own judgements. Linton (1955) was one of the first to test this hypothesis, and in a series of experiments examining the relationship between field dependence on the one hand, and conformity and suggestibility on the other, she found that the overall correlation between conformity and the field dependency was quite substantial (+·60). Similarly, Konstadt and Forman (1965) found that field-dependent children looked at the examiner during a stressful task much more often than field-independent children, presumably for some form of guidance or reassurance. These children were also more affected in their performance by approving or critical comments from the experimenter. According to Witkin's theory the person whose perception is strongly influenced by the physical background (i.e. a field-dependent person) will also be the person whose behaviour is strongly influenced by the social background. The bulk of the evidence available for children and adults would seem to support this.

Assessment
The main measures developed by Witkin and his colleagues are:

The Tilting Room, Tilting Chair (TRTC) apparatus. In the TRTC test, the subject is placed in a chair, which can be tilted to the left or to the right. The chair is situated in a

88

small room, which can also be tilted left or right. The subject is required to control the position of the chair so that his body remains upright despite the tilt of the room. Alternatively, the subject is required to set the room in an upright position when the chair is tilted.

Rod and Frame Test (RFT). In the RFT, the subject is presented with a luminous rod surrounded by a luminous frame in an otherwise completely darkened room. The frame can be tilted through various degrees, and the subject is required to adjust the rod to the upright.

Embedded Figures Test (EFT). The subject's task in this test is to locate a simple figure, previously presented to him, within a larger and more complex design. This design is drawn so that it contains all the elements of the simpler figure, but in such a way that they are obscured. For example, the outline of the original simple figure could also constitute the outline of several other more prominent subfigures within the whole figure.

The stability of these tests is amongst the most impressive in the whole of personality research. For example, Witkin, Goodenough and Karp (1967) re-administered the RFT after intervals of up to fourteen years, and the retest correlation for boys was found to be ·66. For shorter intervals, retest correlations were frequently higher than ·85. A number of studies have found retest reliabilities of this order, and the consistent demonstration of high correlations attests to the genuine durability of the perceptual/cognitive styles measured by these tests.

To the extent that these tests are all intended to measure the concept of differentiation, substantial correlations between them could be expected; here the data is not so impressive. Although it is consistently found that these measures do intercorrelate, the degree of relationship (about ·4) indicates that they can not be regarded as equivalent measures. Vernon (1972) in a factor analytic study, has recently confirmed that the RFT in particular may in part be measuring different factors from the other measures of differentiation. Further-more, some investigators have found only moderate relationships between different commercial versions of the same test.

The foregoing discussion may have suggested that the apparatus used to measure differentiation can be expensive, bulky, time-consuming to administer and cumbersome, and this is in fact so. Accordingly, a number of alternatives have been developed. These include shorter versions of the original perceptual tests and versions of the EFT which can be administered to groups rather than individuals. Such versions do not, however, relate sufficiently highly to the original tasks and to each other to justify their widespread use, the intercorrelations between them being only about ·6. Similarly, one study found no relationship at all between RFT and a questionnaire designed to measure differentiation.

In short, the reliability of the various individual measures of differentiation is high, but their generality is not so impressive. This latter point is particularly important because many investigations into field dependence have used only one measure, as if it could measure all the component aspects of the concept. No single test in itself can do this and, therefore, to use one test tells us little or nothing about field dependence in general, but only about that particular test. An adequate assessment of field dependence would seem to require a battery of tests. Vernon (1972) recommends a portable version of the RFT, a short form of the EFT, and a Draw-a-Person test as the most appropriate constituents of a battery. A combined score from such a battery of tests is called the Perceptual Index.

Applications

Witkin (1965) gives many interesting suggestions concerning the possible utility of field dependence tests in psychopathology and provides supportive evidence where available. Some examples are:

Psychoanalysis. The concept of ego-defence mechanisms is basic to psychoanalytic theory (see Ch. 2). Witkin argues that a relatively undifferentiated person would tend to use primitive crude 'blunderbuss' defences such as repression and denial, whereas more selective defences such as reaction formation would be typical of the more differentiated person. Psychoanalysis could be greatly expedited by early identification of the defences used by the patient. A number of studies in this area have supported these formulations but the findings can not be regarded as conclusive, since the measures used to assess defensive structure were typically 'projective' devices such as the Rorschach and the TAT, whose validity is somewhat open to question (see Ch. 2).

Alcoholism. The global and articulated styles, with their implications regarding self-identity, might be expected to manifest themselves in different ways in various psychiatric disorders. It has often been shown that pathological groups, with some form of dependency as a major component of the pathology, are markedly field-dependent. Alcoholics are often considered to be very 'dependent' (not least upon alcohol) and indeed many studies have shown that the alcoholic is a relatively field-dependent individual. If this conception of the alcoholic as a generally dependent person is correct, it might be predicted that alcoholics would obtain high scores on a variety of field-dependency measures. Goldstein, Neuringer, Reiff and Shelly (1968) tested this hypothesis and, although they provided further evidence that alcoholics are more field-dependent, they were unable to find higher scores with other forms of dependency, such as need for personal support. They concluded that 'some other mechanism other than that proposed by the Witkin group must be postulated in order to explain the nature of the performance of alcoholics on the RFT'. Thus the precise nature of the relationship between field dependency and alcoholism is still unknown.

Psychiatric treatments. Witkin, Lewis and Weil (1968) have produced some preliminary results relating differentiation to patient-therapist interaction during psychotherapy. Some of the differences between differentiated and undifferentiated patients were very marked indeed; for example, the average length of differentiated patients' comments was 158 words, whereas that of undifferentiated patients was only 39. It was also suggested that the degree of differentiation of the therapists was important. Such findings appear to be of major importance in conducting psychotherapeutic sessions, and could clearly have a bearing on the outcome of therapy. If this study can be replicated with larger numbers and more adequate controls, it could constitute an important contribution to psychotherapeutic practice.

Evaluation
Witkin's theory has a great deal of supporting empirical evidence from a wide range of settings. Certainly the size of the correlations with other measures, and the consistency over time, are amongst the highest reported in personality research, and very promising potential for applications in many areas has been indicated. However, Witkin's theory has recently been subjected to a number of criticisms, (e.g. Wachtel, 1972) which will be summarized below.

The most consistently reported correlations of field-dependency measures are with measures of intelligence. Sometimes these correlations are as high as the correlations of field-dependency measures with each other. Indeed, the Draw-a-Person test, a commonly used measure of field dependency, is typically used as an intelligence test. Furthermore, the relationship between the EFT and some subtests of the Wechsler Intelligence Scales is so large that Witkin (1965) recommends the use of such tests when available as adequate substitutes for other field-dependency measures. This immediately raises the question of how far many of the differences found using the field-dependency measures can simply be viewed as indicating differences in intelligence, and have nothing to do with personality measures at all.

There is, of course, no reason why personality measures should not relate highly to other factors such as intelligence, as long as it can be demonstrated that the personality measures add something of value over and above that of the intelligence measures alone. Only two studies on issues central to the concept of field dependency appear to have included adequate controls for intelligence. Minard and Mooney (1969) reported that their results (demonstrating a relationship between the Perceptual Index and recognition of emotionally loaded words) could not be attributed to intellectual factors. On the other hand, Vernon (1972) found that with the effects of intelligence eliminated, field-dependency measures had little or no relationship with a wide range of other personality measures, but some low to moderate relationships with general interests remained. Thus how far many of the findings of the Witkin group can be accounted for simply in terms of intelligence without recourse to concepts, such as differentiation, remains an open question.

Witkin has replied to such criticisms by pointing out that the field-dependency measures correlate chiefly with only a limited number of intellectual measures; in particular, the correlations with verbal items are quite low. Thus, field-dependency tests do not measure overall intelligence at all, but only one aspect of performance on a limited range of tasks, an aspect which both kinds of tests just happen to have in common. However, in contrast, some studies have shown that verbal measures can be quite highly correlated with field-dependency measures. Even if field-dependent and field-independent subjects do differ only on non-verbal tests, then the 'greater total stockpile of intellectual resources could still account for any differences observed' (Wachtel, 1972).

A further difficulty with the differentiation hypothesis lies in the vagueness and ambiguity of many of the concepts used. Few firm predictions can be made from the theory to the extent that almost any finding can be said to be consistent with predictions, thus reducing its scientific value. One might argue that alcoholics (who often drink alone, who continue drinking despite frequent social censure, and who often function very efficiently in circumscribed areas) show many of the characteristics of field *in*dependency. The opposite has consistently been found to be the case, but one wonders whether

93

these findings are explained 'after the fact' rather than being predictable from the theory.

Wachtel (1972) has also argued that a single dimension such as field dependence-independence is too limiting, and that variations in field dependence will manifest themselves in radically different ways depending upon such other factors as overall level of intelligence and other personality factors. A very verbal, fluent extravert above the average on field dependence will probably behave very differently in a number of important ways from a non-fluent introvert with exactly the same score on the Perceptual Index. He argues strongly in favour of an explicit acceptance of the Perceptual Index measures as ability measures, and only indirectly as personality measures. This, he believes, would help to rectify much of the loose thinking in this area, and serve to sharpen the theoretical constructs surrounding field dependence, leading to more specific and more meaningful predictions.

Finally, the meaningfulness of this approach would appear to hinge in part upon the consistency of performance on the Perceptual Index tasks. As mentioned, test-retest correlations tend to be rather high for such tasks, but another way of assessing consistency is to see how far the performance can be modified. McAllister (1970) attempted to change the RFT performance of 60 chronic schizophrenics by operant conditioning, and found that their performance was readily modifiable, and that the changes had continued at follow-up a month later. He concluded that his findings 'open the possibility ... that these aspects of individual functioning are more a reflection of similar kinds of life experiences and histories than of internal factors comprising personality structure.'

To conclude: in the area of personality theory, Witkin and his associates have consistently produced some of the highest correlations with respect to reliability and validity. It is precisely the magnitude of these relations that makes one suspect that we are not dealing with personality factors at all, but with cognitive, intellectual factors. It may be most useful to regard Witkin's measures as intellectual measures; if we do, many of the findings on field dependence and related constructs fall into place. In particular, the findings of Witkin *et al.* (1967) on the sequence of developmental stages of field dependence from infancy, and the suggested constitutional determinants, are entirely consistent with viewing the Perceptual Index measures as indirect measures of intellectual ability.

However, even if this is the case, a great deal of credit remains to be given to Witkin. His work has thrown new light upon and raised new issues concerning exactly what standard tests of intelligence measure. In short, neither viewing the concepts related to differentiation as personality variables, nor as intellectual variables, gives a completely satisfactory account of the results to date. Further research, and delineation of variables will be required before a solution to these problems can be offered.

6
Narrow band
theories (2)

(1) *Need for achievement*

Traditionally, the psychologist's approach to human motivation has been to compile a list of needs (see D2); although this general approach is now largely discredited, some of the needs that have been specified still exert a major influence on current psychological theory. The psychological needs outlined by Henry Murray (1938) have been particularly influential; from them has developed a number of widely-used personality tests and a great deal of research. By far the most intensively investigated is the *need for achievement* (or n Ach as it is commonly referred to); much of the work in this area has been conducted by McClelland and Atkinson, and their colleagues in the United States.

N Ach is concerned with accomplishment, and how people vary in the degree that they are prepared to strive to succeed in competition towards a particular standard of excellence. It is related to such things as overcoming difficulties; maintaining high standards and improving one's own performance; competing with the level established by others, and generally gaining mastery over one's physical and social environment. A person with high n Ach would tend to indulge in tasks which would tax his competencies. Tasks that are too easy would be comparatively unattractive, since, with little chance of failure, there is no real challenge. Similarly, extremely difficult tasks would be avoided, offering as they do little chance of success.

Independently of their need for achievement level, people can also vary in their motivation to avoid failure. High 'Fear of Failure' or FF subjects will tend to lack self-confidence and generally have a poor opinion of their capabilities. They will indulge in easy tasks in which there is little chance of failing, or paradoxically in extremely difficult tasks where their failure is most easily accounted for in terms of the inherent difficulty of the task, rather than any personal deficiencies. An example is that of a climber who consistently tries and fails to climb the most difficult routes. He receives much social approval for his attempts, and his failures are seen as resulting from the extremely difficult nature of the climbs, rather than his poor climbing abilities.

Atkinson and his colleagues (for example, Atkinson and Feather, 1966) have developed this theory, regarding n Ach as a motive that, when seen in combination with other variables, can lead to predictions of behaviour in a variety of settings. The three most important factors bearing upon a person's success will be: his degree of achievement motivation, his fear of failure and the incentive or the value of the accomplishment to the individual. Atkinson and Feather have developed a series of mathematical models describing the relationships between these three variables. From these models, many predictions have been made which have been supported in empirical tests; for example, subjects who have higher need for achievement than need for avoiding failure tend to choose occupational goals compatible with their abilities. Subjects with the opposite pattern choose occupations that are relatively very easy or very difficult for them, given their level of ability.

It should be emphasized that McClelland, Atkinson and their co-workers do not assume n Ach and related motives to be genetically determined. Such motives are considered to be learned, mostly in childhood. Children who are reinforced

97

for accomplishments will tend to be high n Ach scorers, and children punished for failure will tend to have high FF. Nevertheless, although these motives are learned in childhood, they can be developed or unlearned throughout life.

McClelland, Atkinson, Clark and Lowell (1953) in their book *The Achievement Motive*, described a series of studies demonstrating the importance of n Ach in a wide range of settings – students' examination marks, persistence at anagrams, performance on arithmetic and verbal tasks, effects of failure and conformity, to name but a few. All of these provide some support for the validity of the concept of n Ach. Since then, the relationship between achievement motivation and economic advancement has been intensively investigated.

A number of studies have demonstrated a link between n Ach and social mobility. For example, one study found that high n Ach persons were more likely to have risen in occupational and social status; whereas low n Ach persons tended to remain at the same level or to move in a downward direction.

The scoring criteria used to assess n Ach from phantasy material (see p. 99) can be similarly applied to assess the overall level of n Ach in a culture, from its myths, pottery, children's readers, short stories and so on. In *The Achieving Society*, McClelland (1961) has indicated, amongst other things, that: tribes whose folk tales reveal high n Ach are likely to be characterized by higher levels of entrepreneurship, such as trading and craft work, than are low n Ach tribes; that there is a strong relationship between n Ach in England from 1500 to 1800 (as measured by achievement themes in English literature), and economic activity (as measured by coal import figures for London); and that the vicissitudes of the ancient Greek civilizations were preceded by parallel fluctuations in achievement motivation (as measured by literary themes).

Contemporary cultures and civilizations have also been subjected to such scrutiny; measuring n Ach by themes appearing in children's reading books, and economic growth rate by increases in the production of electricity, McClelland has demonstrated a moderately high correlation of +·43 between them, collecting his information from countries from all over the world. Furthermore, there seems to be a relationship between n Ach and climate, such that the peak of n Ach occurs in climates with an average temperature of between 40°F and 60°F; deviations from this range in either direction are accompanied by a decrease in n Ach. Finally, McClelland has noted a relationship between Protestantism and n Ach, with Protestant countries tending to be more economically advanced than Catholic countries.

In many of these historical and cultural studies, there seems to be a time-lag of around fifty years between the development of high n Ach levels and actual increases in economic activity, and the idea has been put forward that levels of n Ach may be used to predict economic level five or so decades later. The finding of a fairly regular decline in achievement motivation in American literature from the early 1900s to the 1960s is of some import. Whether this is desirable does of course depend on the value one attaches to economic advancement.

Although not all research findings have been positive, the range and depth of the sources of support for the theory are particularly impressive in helping to establish the validity of n Ach as an important personality variable.

Assessment of n Ach and FF

The original and still the most frequently used method of obtaining n Ach measures was through the ratings of phantasies of achievement expressed in response to TAT cards (see p. 29). Generally, the reliability of the TAT (in terms of stability over time) is quite low; nevertheless, McClelland *et al.* (1953) suggested some specific criteria for judging achievement phantasy on the TAT, and acceptably high inter-rater reliabilities were obtained, of about ·9. Administering the TAT to the same group of subjects at times separated by various intervals up to six months, we get somewhat smaller

99

reliabilities, normally around ·2 or ·3. It would seem that although TAT measures of n Ach can be reliably scored in terms of agreement amongst judges, test-retest reliability is mostly unacceptably low.

Accordingly, many attempts have been made to develop alternative measures, some of which are also based on projective tests, others on questionnaires. Weinstein (1969) examined the inter-relationships between eight different measures of n Ach and FF (three projective and five questionnaire) and found them to be extremely low. He was forced to conclude that the various devices were not measuring the same thing. Unfortunately, many studies have regarded these different measures as being largely inter-changeable which they clearly are not. To the extent that reliability imposes restrictions on validity and utility, these findings must raise serious questions concerning much of the published research on achievement motivation.

On the other hand, it could be argued that what is of importance in achievement motivation research is not the consistency of a person's specific score on a test, but simply whether, over long periods of time, he remains a high achiever or a low achiever. To use specific scores along a continuum in the assessment of reliability may be inappropriate with a classification system which is fundamentally binary (that is, high n Ach versus low n Ach). Many studies have demonstrated that broad achievement dispositions are reasonably constant from childhood to early adulthood, and from adolescence to the late thirties. Nevertheless, the measurement of n Ach is beset with difficulties.

Applications of achievement motivation
The theory of achievement motivation has been applied in a variety of areas; two have been selected for discussion, management training and education.

Management Training. McClelland and his co-workers maintain that levels of n Ach can be raised by appropriate training and that sufficient training applied to a wide range of people

100

could promote overall economic advancement in any given society. The most ambitious test of this hypothesis is that of McClelland and Winter (1969). Their original, highly ambitious plan (which included a large-scale controlled trial of the effects of n Ach training in several industries in India, Southern Italy and a massive project in Tunisia) had to be curtailed quite drastically because of lack of supporting funds. Their final project, although very interesting and valuable, consisted mainly of an examination of achievement training with about eighty businessmen in two small cities in Southern India. The training consisted of instruction in setting and attaining standards of excellence; taking personal responsibility; discussing personal motivations, opportunities and goals; group discussions of progress, and so forth. The businessmen were followed up over two years to assess any changes in entrepreneurial activity, and compared to a similar group who had had no achievement training. The trainees had made more profit, employed more labour, invested more money, and worked longer hours than the control group. Similar programmes have since been established in a number of areas in need of economic development; most have been equally successful. The importance of this for under-developed countries needs no elaboration here.

Education. Several attempts have been made to use achievement training to foster educational progress. Kolb (1965), for example, used achievement training in a project designed to produce higher grades in underachieving boys. Compared to boys who had no such training, their marks were significantly higher and the superiority lasted for up to two years afterwards.

The main findings of the effects of achievement training in education have recently been summarized by McClelland (1972). The chief conclusions were that for maximum effectiveness, achievement motivation training should be integrated fully in the general school curriculum, rather than being offered in brief intensive courses. Furthermore, the training should be done by everyday teachers who have in turn re-

101

ceived achievement training. The 'climate' of the classroom should be such that it encourages initiative and self-reliance; the materials used should be novel and varied, so that the pupils pay attention to them. Methods should be tailored for individuals, and immediate feedback should be given concerning the pupil's performance.

From the above, it will be obvious that the recommendations are more and more concerned with general classroom management skills, the relevance of achievement training being somewhat remote. McClelland acknowledges this, and considers that it may be possible that 'achievement motivation training is effective in the classroom without much affecting the level of achievement motivation in the students'. The way in which achievement motivation training produces effects in the classroom is indirectly, by improving the teacher's classroom management techniques, which in turn promotes more attention, participation, and responsibility from the pupils.

Evaluation

The theories put forward by McClelland and Atkinson are exemplary in terms of breadth of vision, imagination and creative conception. Their efforts are particularly commendable in that they started with a theory, tested it in some detail in the laboratory, derived mathematical formulae to describe it, examined it in real-life situations, and then most importantly they analysed its utility and attempted to apply it to important issues such as economic advancement in poor nations, and education. From a theoretical viewpoint it is also noteworthy that they are willing to achnowledge the limitations of their work, and to see it in the context of other processes.

Nevertheless, there are some problems associated with the theory of achievement motivation, in addition to those measurement problems referred to earlier. Hurley (1971), for example, has highlighted some of the sampling and statistical difficulties that often characterize research in the area of achievement motivation training. There are also many theoretical issues to which answers are not apparent; for example,

102

is a person who is highly motivated to do well in an occupational setting similarly motivated to perform well in a sporting, social or general economic setting? If not, why do some people channel their n Ach in particular directions rather than in others?

Much of the research on n Ach is simply concerned with establishing relationships between achievement and other variables, such as educational attainment, economic progress and so on. McClelland and his associates assume that changes in n Ach *cause* changes in these other variables but there are clearly many other possible interpretations. For example, the relationship between social mobility and n Ach may result from social climbers becoming increasingly preoccupied with achievement and this is reflected in their phantasies and, therefore, in their n Ach measures. Many of the cross-cultural and historical investigations of the correlates of n Ach can be interpreted in the same way.

Perhaps the most telling criticism is that the changes consequent upon achievement motivation training may not be due to the training *per se,* but to other factors operating simultaneously. In the Indian studies, for example, the trainees were given an enormous amount of attention from world-renowned figures, they were instructed in how to set up and develop a business, they formed many new business acquaintances, and they were given more financial backing than they would otherwise have got. Any one or combination of these factors could account for the changes. This might well be the most appropriate explanation of the changes found, in that over the two or three years after training, the Indian businessmen demonstrated a decline in n Ach scores; this decline occurred whether or not they became successful businessmen. Thus the need for achievement which was augmented by the training had no necessary or enduring relationship to their business acumen. Successful trainees were already very highly motivated before the training, and the training may simply have directed their motivation into useful and productive channels. Similarly, analysis of the results of Kolb's (1965) study indicates that the academic improvement in the under-

103

achieving boys was largely restricted to the boys from relatively high socio-economic backgrounds. It could be argued that such a background would already provide the boys with some sort of achievement training and with a sympathetic and responsive environment, conducive to the expression of high need for achievement.

(2) *Sensation-seeking*

Over many years laboratory evidence has accumulated, indicating that a wide range of organisms can be trained to perform various tasks in order to produce varied experience and stimulation, which have no immediate connection with any form of biological needs or drives. Such stimulation seems to be sought out for its own sake. For example, Butler (1958) indicated that monkeys could be trained to perform complex tasks when the only consequence was to see such things as another monkey or a moving electric train. Similar results have been demonstrated with mice, rats, chimpanzees and human children and adults. This 'Sensory Reinforcement' or 'Sensation-Seeking' seems to function in a very similar way to primary reinforcement, such as food and drink, and may be of fundamental biological and psychological significance (Kish, 1966).

There is no doubt that people vary greatly in their preferred level of stimulation, and many psychologists have hypothesized a continuum of sensation-seeking tendency. The high sensation-seeker would typically seek out novel, unusual, complex and unpredictable situations, whereas the low sensation-seeker would prefer a quiet, unvaried, straightforward, routine environment. If there are any marked changes in stimulation, the person will act to restore the balance in the direction of his preferred level, for example, by adjusting the volume of a radio or by exposing himself to a dangerous situation.

Recent interest in sensation-seeking has resulted from experiments on *sensory deprivation*. This general term covers a variety of experimental procedures, designed to investigate in humans the effects of being subjected to minimal levels of stimulation. For example, subjects might be placed in a dark, sound-attenuated room, with long padded cuffs over arms and hands, and with temperature held constant. The amount of in-coming sensory information is, therefore, severely restricted. Many interesting results have been found in these experiments; for example, under some conditions, subjects reported very bizarre experiences such as hallucinations and extreme panic (see Zubek, 1969, for a comprehensive review). People who have differing preferred levels of stimulation could clearly be expected to differ in their tolerance for sensory deprivation situations. A leading research worker in the area of sensory deprivation is Zuckerman, and he devised a Sensation-Seeking Scale in this connection (see later).

More recently, Zuckerman and his colleagues have suggested that sensation-seeking is a basic personality dimension that reveals itself in many different situations. Knowledge of a person's score on the Sensation-Seeking Scale should enable us to predict behaviour, not just in relation to sensory deprivation, but in a whole range of situations in which the tendency to seek or avoid sensation could be an important component. Zuckerman is not alone in considering sensation-seeking to be of some importance for human personality. The concept has played a major role in many other theories of personality; for example, Eysenck suggests that individuals high on extraversion and psychoticism tend to be high sensation-seekers. From many studies examining the personality of high sensation-seekers, a picture emerges of them as being more creative and intelligent, better adjusted, independent, unconventional and impulsive. In addition, Zuckerman and his colleagues have recently put forward some interesting speculative ideas concerning the relationship between sensation-seeking and the overall level of arousal (or alertness) in the brain. Preliminary findings suggest that there may be some important biological differences between high and low sensation-seekers.

Assessment of sensation-seeking

Many tests have incorporated items relating to sensation-seeking, but only those relatively 'pure' measures of sensation-seeking will be mentioned here.

Sensation-Seeking Scale (SSS). Originally developed by Zuckerman *et al.* (1964) in connection with research on sensory deprivation, this is the most widely used measure. The original form has 34 two-statement items, and the subject is required to choose which of the alternatives is most appropriate for him; for example,

11 A : I sometimes like to do things that are a little frightening

B : A sensible person avoids activities that are dangerous

26 A : I prefer friends who are excitingly unpredictable

B : I prefer friends who are reliable and predictable

Reliabilities of the order of ·7 to ·8 have been consistently reported and evidence of validity has been provided chiefly by correlations with related scales. A number of improvements have been included in later versions of the SSS. Zuckerman (1971) has recently produced Form IV, a 72-item version which has four factors in addition to the general factor: Thrill and Adventure Seeking, Experience Seeking, Disinhibition, and Boredom Susceptibility. Thrill and Adventure Seeking refers to a desire to participate in outdoor activities that involve speed or danger (parachuting or fast driving). Experience Seeking refers to the wearing of flamboyant clothes, behaving 'wildly', experimenting with drugs, preferring unusual forms of art and music, and a general restlessness and seeking out of new ideas. Disinhibition refers to hedonist behaviour, such as heavy social drinking, varied sexual activities, and heavy gambling. Finally, Boredom Susceptibility refers to a dislike of monotony and routines, and an avoidance of dull people or situations. Reliabilities for this version have been reported to be in the same range as the earlier scales, both for the general score, and for the factor scores.

Other measures have also been devised (e.g. the Change

Seeker Index and the Stimulus Variation Seeking Scale); these tests correlate with each other and with the SSS at about $+ \cdot 66$, and they are, therefore, to some extent measuring the same variable. A very promising new test is the Arousal Seeking Tendency (Mehrabian and Russell, 1973), but none of these other measures have as yet been sufficiently investigated to warrant detailed discussion here.

Applications

The SSS was originally developed as a method of predicting response to sensory deprivation. If people with a high need for stimulation are placed in a situation whereby stimulation is radically reduced, it could be predicted that their tolerance for the deprivation condition would be lower than that of people with a lower need for stimulation. On the whole, however, no consistent relationship has been found between SSS scores and tolerance for deprivation (see Zuckerman *et al.*, 1967).

It is possible that the SSS could be of value in helping to diagnose pathological conditions but findings are again inconsistent. Schizophrenic patients have been reported to have both higher scores in some studies and lower scores in others; where significant differences between schizophrenics and other kinds of patients have been found, they tend to be of small magnitude. Kish (1970) found only weak relationships between SSS score and whether the subject was schizophrenic, alcoholic, or normal. Similarly, no consistent differences have been found between psychopaths and non-psychopaths, despite expectations that psychopaths would be high sensation-seekers (Blackburn, 1969).

A prediction from theories of 'optimal level of stimulus input' is that individuals will vary in the preference for designs, with high sensation-seekers preferring complex, varied designs, but low sensation-seekers preferring greater simplicity. Many studies have investigated this problem, but the results have been somewhat inconsistent. Looft and Baranowski (1971) for example, found moderately low correlations (about $+ \cdot 33$) between three different sensation-seeking questionnaires and preference for complexity in abstract shapes, and they concluded that 'Pencil-and-paper tests appear to be measuring one kind of trait, and the random-shapes method seem to be tapping quite another trait.'

Zuckerman *et al.* (1972) report on a number of studies in which SSS scores were related to specially constructed scales of sexual experience. The latter consisted of items assessing a wide range of sexual behaviours including intercourse in various positions, oral-genital contacts, number of sexual partners, and so on. The main finding for men was that high sensation-seekers had experienced a wider range of sexual activities, this applying to all the sub-scales within the SSS. For women, however, the major correlation of sexual experience was with the Disinhibition Scale, and it was considered that indulgence in a variety of sexual experiences may be part of the general sensation-seeking trait in males, but is a more specific aspect of sensation-seeking in females.

A major reason for the non-medical use of drugs is to increase stimulation and arousal, although the motivation for drug-taking varies greatly from person to person. In general, it could be hypothesized that drug-takers would have higher SSS scores, and indeed this seems to be the case (Zuckerman *et al.*, 1972). As one might expect, this applies particularly to the 'cortically-arousing' drugs, such as amphetamines, and LSD, and somewhat less to 'cortically-depressing' drugs, such as heroin; these 'depressant' drugs seem to be preferred by the low sensation-seeker. With more socially accepted drugs like alcohol and nicotine, the relationship is much less clear. Information on the characteristics and motivation of people who take drugs of various kinds, could well be of value in arranging appropriate treatments where necessary.

Evaluation

The theory of sensation-seeking put forward by Zuckerman and others seemed likely to prove fruitful for a number of reasons. First, sensation-seeking (or sensory reinforcement) has been extensively investigated in the laboratory, and there is, therefore, a large body of experimental knowledge underpinning the theory. Second, the theory has clear links with recent developments concerning arousal and the brain. Third, experiments, particularly with animals, indicate that sensation-seeking may be of fundamental psychological and biological importance. Finally, the importance of sensation-seeking is widely acknowledge and although theorists may differ in the details of their formulations, such consensus of opinion is encouraging.

The results of much of the research in sensation-seeking have, however, been found disappointing; contradictory results are common, and where consistent results from study to study have been found, the relationships between sensation-seeking and other variables tend to be rather weak. In addition, some unexpected findings have emerged which cast doubt on the validity of the sensation-seeking variable; for example, one study found marked differences in sensation-seeking scores between two groups of students when the only apparent difference between the student groups was that they were from different universities.

A major shortcoming with many of the validity studies on sensation-seeking is that most have related sensation-seeking only to other personality tests. Although this procedure can add support to the validity, it is not without its problems, particularly when attempts are made to replicate the findings. For example, Zuckerman and Link (1968) found many significant relationships between the SSS and a number of personality measures; when repeated by Waters and Waters (1969), however, many fewer relationships were found, and even the ones found were much weaker. Some of these dangers may be avoided by relating SSS to actual behaviour in natural settings; this has unfortunately seldom been done.

It is to be hoped that the use of the more recent scales, measuring the various components of sensation-seeking rather than regarding it as a single trait, will produce more consistent results and add to our knowledge of this potentially valuable area.

A critical overview of the trait/type approach

Many psychologists have expressed dissatisfaction with the notion of personality traits. Foremost amongst these critics is Mischel, whose book *Personality and Assessment* (1968) has been extremely influential. The arguments of Mischel and other critics are based on theoretical and empirical grounds, and will be discussed below.

A serious theoretical objection is that a trait approach pays little heed to the evolving process in human personality. It takes a static view, in that the person is examined at one particular moment in time, but as Holzman (1974) states: 'personality never *is,* it is always *becoming*'.

Empirical objections have also been raised. Trait theories rest fairly and squarely on the explicit assumption that there are cross-situational consistencies in behaviour; that being an introvert, for example, means that the introversion will manifest itself in many situations. According to this view, personality consists of broad dispositions emanating from within the individual which exert a powerful influence on a person's behaviour across many situations, and over time.

However, little empirical evidence of consistency of personality has been found. If one compares questionnaire responses with assessments obtained by other techniques (such as simple observation) the relationship is normally trivial; and the degree of consistency drops dramatically when the same person is assessed over time but in different situations.

Perhaps a few examples will help to clarify this point. The earliest and most frequently quoted study bearing on this issue is that of Hartshorne and May (1928). They investigated 'moral character' or honesty in schoolchildren in a wide variety of situations and found very little consistency. Some children who were honest in one situation were dishonest in others, and situational effects seemed to exert a greater influence than the children's 'sense of honesty'. Much work has been done in this area since, but the bulk of it does not necessitate a revision of Hartshorne and May's conclusions that such things as honesty, deception, persistence and so on are 'groups of specific habits rather than traits' (e.g. McGeorge, 1974). Under close scrutiny, traits are inevitably found wanting in terms of cross-situational consistency and this applies even to such specific traits as curiosity, dependency, punctuality, risk-taking, rigidity and conditionability.

If a person's behaviour is not consistent across situations, we would not expect accurate predictions of behaviour, either from scores on a test or from behaviour in another situation (unless the situations are very similar). Mischel has used the term 'personality co-efficient' to describe how far scores from a questionnaire can be used to predict behaviour in another situation. Reviewing many studies, Mischel (1968) has found that the typical correlations are in the range of ·2 or ·3. These low correlations raise serious questions for those who maintain that personality traits are a major source of individual behaviour.

Correlations can be converted (by simply squaring the co-efficient) into the statistical concept of 'proportion of variance accounted for'; this figure gives us an approximate idea of how much of the variation amongst a group of scores can be said to be due to or explained by the score on the related measure (see A9).

If, therefore, we square the typical personality co-efficient, we find that personality traits (as measured by the relevant personality test) account on average for less than ten per cent ($·3^2$) of the variation in behaviour in other situations. Serious errors may result if we make important decisions about people on the basis of measures with such weak predictive power.

Mischel (1972) reported that 'the predictions possible from a subject's own simple direct self-ratings and self-reports generally have not been exceeded by those obtained from more indirect, costly and sophisticated personality tests, from combined batteries, and from expert clinical judgements'. Furthermore, just looking at previous performance in similar situations may provide us with equally accurate predictions. This conclusion has been shown to apply to predictions in a wide variety of practical situations, such as college achievement, professional success, outcome in psychiatric treatment,

111

parole violations, vocational guidance, and even suitability for working in the Antarctic.

Apart from this lack of supporting evidence for consistency, do we have any direct evidence for the notion of inconsistency? There is indeed much evidence bearing on this issue. A recent study by Allen and Potkay (1974) will serve to illustrate.

Allen and Potkay (1974) examined the variability of self-descriptions in a group of twenty-six university students over thirty-seven days. The students were asked 'Please write five adjectives that describe how you feel about yourself today ... Consider how you feel about yourself on this particular day. You need not reflect on how you have felt about yourself on past days.' The favourability of the adjectives generated was rated and averaged for all the individual students across all days. Without exception subjects wrote down adjectives which were very favourable on some days, but were very unfavourable on others. The self-assessments fluctuated within a very wide range, subjects often reporting both positive and negative evaluations on the same day. As the authors noted, the results 'may be used to question some traditional conceptions about an individual's being rather constant over time in his perceptions of himself, or in his feelings, or descriptive ideas about himself. The results also question common assumptions that an individual deviates from his self-descriptive norm only after extraordinarily special or traumatic events. Favourability of self-description changed on virtually a day-to-day basis.' It would appear then that with self-perceptions inconsistency rather than consistency is the more usual circumstance.

Traits, therefore, do not show consistency over time or across situations; they enable only weak predictions to be made, at best, and often the predictions are less accurate than those based on 'common sense'. But the notion of traits still persists. Why?

First, we tend to have 'implicit personality theories', or our own ideas of what personality traits are typically found together (Schneider, 1973). Our theories are vague and not subject to disconfirmation, and if we observe behaviours not

112

consistent with our implicit theories, we tend to rationalize and force them to fit, perhaps by manufacturing additional information. Furthermore, we tend to see people only in a very limited range of situations. Our initial impressions are very strong and resistant to change and even wildly inconsistent behaviours can be brought into a consistent view. It is also possible that people seek out particular situations, and by their own behaviour set up consistent social environments, which in turn generate consistent behaviours.

Jones and Nisbett (1971) provide a further clue; they suggest that when judging others, we tend to look for and find the causes of their behaviour in stable personality traits; but when we judge our own behaviour, we tend to see it more as a result of situational factors. Others behave in particular ways because of their personality characteristics; we behave in particular ways because we decided to, or because we are tired and so on. We have a more detailed knowledge of our own circumstances, our previous experiences, and current state, and so have less need to attribute our behaviour to unobservable factors such as traits.

Finally, some characteristics do show consistency, such as intellectual and cognitive abilities, and appearance, and we may overgeneralize from this to conclude that all behaviours are consistent.

To conclude, it would appear that personality traits, like beauty and the retina, are in the eye of the beholder.

7
A social learning approach to personality

The social learning approach to personality has derived its main impetus from the 'behavioural' movement in psychology and partly from a dissatisfaction with other, particularly trait, approaches.

Behavioural psychology
There is no single 'behavioural' position in modern psychology, the approach embracing widely disparate views. Some behaviourists maintain that the main source of behaviour lies in the environment, and not within the person. Factors such as genetic influences or dispositions from within the person are not neglected, but they are judged to be of minor importance. References to personality traits, physiological mechanisms, what goes on 'under the skin', or 'inside the skull' are avoided and 'mentalistic' explanations of behaviour, such as those posited by Freudians, are seen as unhelpful and even misleading. Rather, a person develops and behaves in accordance with what happens to him throughout life, or what he *learns*. Behaviour is potentially explicable in terms of environmental effects and changes in the environment can bring about changes in behaviour.

A leading exponent of this view is B. F. Skinner (see Skinner, 1974), a major influential figure in modern psychology.

114

His experiments in the area of *operant conditioning* (or the conditioning of voluntary behaviours) have been accompanied by a systematic philosophy of human behaviour and of science, all of which have had a profound influence in many areas within psychology and other fields. However, Skinner has seldom directed his attention specifically to the field of personality, but his approach and research findings have played a central role in most behavioural personality theories. Clearly the implications of a Skinnerian view are profound when applied to the study of human behaviour, and of personality in particular. For an account of operant conditioning, see A3; for a review of its applicability to social behaviour, see B1; and for critiques of its underlying assumptions, see F1 and F8.

However, when a basically Skinnerian viewpoint has been adopted, many personality theorists have found it necessary to incorporate some 'inner' or cognitive variables to account for particular aspects of human behaviour, as we shall see. But before describing these developments, it is necessary to clarify the use of the terms *'learning'* and *'environment'*.

Psychologists use the term 'learning' (or conditioning) in a very broad sense, to cover all cases of a change in a person's behaviour as a result of his experiences, from learning a skill such as driving to learning the standards of behaviour expected by society. Essentially the same principles are assumed to underlie the learning of all behaviour. A person's 'environment' refers to more than simply his physical, social or geographical background, but would also include other people and their behaviour, in fact every thing and every event external to the person.

Conditioning is a vast and ever-expanding area of investigation and a large body of knowledge has accumulated, but it is impossible to cover the complexities and extent of this knowledge here (for an extended discussion see Blackman, 1974). However, some reference to the basic principles is necessary.

A main type of conditioning is 'operant' conditioning. Although this had been investigated since the early twentieth century, it received little attention until several years after the publication of B. F. Skinner's *The Behaviour of Organisms* (1938). Operant conditioning is concerned with the conditioning of operants (i.e. responses that operate on the environment). These are basically voluntary (non-reflex) behaviours and it is alleged that this type of conditioning is, therefore, of particular relevance to human behaviour. With operant conditioning the emphasis is on responses and the animal must do something or respond before conditioning can occur.

The typical experimental set-up with operant conditioning is as follows: a pigeon is placed in a cage which contains a disc on one of the walls, and a food tray; on the outside of the cage is a food dispenser which is activated whenever the disc is pecked. A peck on the disc is, therefore, soon followed by the arrival of a pellet of food in the food tray. The arrival of the food is said to strengthen or reinforce the peck at the disc; that is, *reinforcement* (arrival of food) increases the likelihood that the pigeon will again peck the disc. If pecks at the disc are regularly and immediately followed by reinforcement, then the pigeon will soon be pecking the disc very frequently. This process of positive reinforcement is a basic mechanism whereby changes in behaviour can occur.

If we then deactivate the food dispenser, so that responses are no longer followed by reinforcement, the animal will gradually cease to respond; this process is called *extinction*. A second major way in which the animal's responding can be suppressed or eliminated is by the use of an aversive procedure in which a response is immediately followed by, for example, an electric shock. This is technically known as a *punishment* procedure

What is learned in one situation does not always generalize completely to another. Some generalization can occur if the subject is placed in a similar situation to that in which the original learning occurred; but, the more the situations are different, the less is the transfer. Much behaviour is, therefore, largely specific to the situation in which it was learned, and the situation exerts a controlling influence over the occurrence

116

of the behaviour. Some situation variables are very powerful determinants of human behaviour, as we shall see.

These principles of learning, which were mainly derived from laboratory experiments with animals, seem to be valid for many different species, including man. However, these principles alone do not give a satisfactory account of the complexities of human behaviour. For example, there is extensive evidence that people do not have to experience events personally or directly in order to learn, and that much human learning takes place at an observational level (Bandura, 1969). In other words, we observe what others do, and note the consequences of their behaviour. If the behaviour is reinforced, we are likely to imitate it, or model our own behaviour on that of the observed; if the behaviour is punished, we are unlikely to imitate it. The work of Bandura and his colleagues is of central importance in a social learning approach to human behaviour.

In sum, from a behavioural viewpoint, to look for the causes of behaviour within the person is to look in the wrong place; trait theories divert our attention from the important controlling variables in the environment. What seems to be important is what a person *does*, not what he *has*. Learning experiences are unique for each individual and individuals will, therefore, behave in unique ways, and not in accordance with traits.

Situational determinants of behaviour
Attempts to account for human behaviour by reference to person variables have been found wanting. Does an account in terms of situational variables appear more promising? It is clear that the same situations can evoke very different behaviours in different people, and that seemingly different situations can evoke similar behaviours in one person. Thus, it is not situations as such that evoke behaviour, but how a person construes them. Situations have a different 'meaning' for each person, and will exert a varying influence, depending on his previous learning experiences in those situations.

People do not passively and indiscriminately admit all out-

117

side influences, but select and evaluate them. This selection and evaluation has a marked effect on how a particular stimulus affects behaviour. Mischel (1973), for example, discusses compelling evidence to this effect. He placed children in front of a selection of attractive sweets and told them to refrain from eating them until the end of a specified interval; otherwise they would be given a more immediate but less preferred object. It was found that how long a child resisted temptation could be readily manipulated. If the experimenter drew the child's attention to the taste or 'crunchiness' of the sweets, the child was unable to resist for long. But if instead, the same child was told to imagine the sweets as being 'little brown logs' or 'round white clouds', he could wait much longer. As Mischel states 'The results clearly show that what is in the children's heads – not what is physically in front of them – determines their ability to delay.' We do not approach each situation without the aid of previous learning (either directly or through observation) in similar situations. Situational factors are not, therefore, sufficiently powerful in themselves to account for human behaviour.

An interactionist solution
We have been discussing the question of the important determinants of behaviour and it seems that neither person variables nor situation variables can in themselves provide a satisfactory account. It may be that the problem expressed this way is meaningless and unanswerable, analogous to asking 'Which is the more important in a car – wheels or the engine?' Looked at in this way, we can see that the importance lies not in one or the other, but in the interaction between them, or how they work together and under what circumstances.

Bowers (1973) has recently reviewed eleven studies that he was able to find which allow one to partition the controlling variables over behaviour into person, situation and interaction variables. The behaviours examined covered a wide range and included aggression in young boys, anxiety in students, and resistance to temptation in children. On average thirteen per cent of the variance was due to person variables, and about

118

ten per cent was due to situation variables, but the inter-action between them accounted for about twice as much (twenty-one per cent). The source of the remaining variance could not be clearly attributed to any particular variable and was assumed to be due to chance fluctuations or 'error'. The above percentages should not be taken as a reliable indication of the actual proportion of influence due to these variables, but are probably most appropriately viewed as guides to *comparative* influence. In other words, the interaction may account for twice as much variance as either persons or situations alone. The actual proportion of the total influence due to these variables may be considerably higher.

Bowers (1973) and others have also pointed out that the proportion of variance due to person, situation or interaction variables will depend on the particular behaviour being con-sidered, and that the relative influences are only *average* in-fluences. In one study he reviewed, for example, concerned with smoking, forty-two per cent of the variance was due to person variables, whereas only seven per cent was due to situations; on the other hand, with talking, sixty-eight per cent was due to situations, but only ten per cent to persons.

In sum, the evidence is not compatible with either a com-pletely situationist position nor a completely trait position. Any theory will have to take simultaneous account of both in-fluences if it is to provide useful predictions about individual behaviour.

Person variables
Conventional notions of personality traits have failed to pro-vide us with an adequate account of person variables. But common sense would indicate that people do have consistent characteristics, so why have several decades of intensive re-search failed to illuminate the precise nature of these charac-teristics? Mischel (1973) suggests some possible reasons, and outlines a number of person variables as alternatives to traits, placing much emphasis on cognitive variables.

A possible answer was also suggested several decades ago.

119

Allport (1937) put forward that traits do exist but that they are idiosyncratically organized in each person. Many theorists class specific behaviours into larger conglomerates, and for some behaviours this may be a valid procedure. But there is no reason to assume that the same specific behaviours will necessarily conglomerate in the same way for all people for all time.

Preference for spicy, exotic foods may be one man's sensation-seeking, but another's oral fixation. A specific behaviour or group of behaviours may form part of any number of traits; for example, the concept of risk-taking figures prominently in many of the trait theories referred to in this book, including Locus of Control, Extraversion, Need for Achievement, and Sensation-Seeking; each theorist seems to have 'adopted' risk-taking as being highly relevant to the traits contained in his own theory. This is despite the fact that risk-taking itself does not seem to be a consistent characteristic of individuals (Weinstein, 1969). On the other hand, some specific behaviours may be appropriate only for a single trait. Punching someone in the face will almost unanimously be seen mainly as part of a trait of aggression, for example.

The position adopted here is not that traits do not exist, but that *universal* traits do not exist. People construct, on the basis of their learning experience, their own structure of behaviours which they group into traits, and a core or dominant trait for one person may not even be acknowledged to exist by another. This conceptualization goes some way to account for the inability of trait descriptions to predict specific behaviours, and also helps to explain the frequently found 'factor variance' (see Ch. 4).

Kelly's Personal Construct Approach (see Ch. 3) adopts a similar position. It will be remembered that in the Repertory Grid Test the subject is allowed to generate his own conceptual scheme to describe himself and his relationships. The subject himself is allowed to decide which specific behaviours form part of any particular trait.

Bem and Allen (1974) have explicitly tested these ideas. They hypothesized that people who see themselves as consistent on a particular trait will in fact be more consistent across a number of situations than people who see themselves as being variable on that trait. The subjects were asked 'How much do you vary from one situation to another in how friendly and outgoing you are?' and '...in how conscientious you are?' and to rate themselves on these traits in a number of standard social situations. They were allowed to use their own concept of the trait, and to exclude those situations that they did not consider to be relevant. The traits of friendliness and conscientiousness were also measured by a variety of other methods, including parents' reports, peer's reports and assessments of friendly and conscientious behaviour in real-life settings.

They found that the correlations between the measures for those who saw themselves as consistent were substantially higher than for the high variability subjects, the average correlations being +·57 and +·27 respectively. With the consistent subjects on the trait of friendliness, correlations across situations as high as ·75 were reported and similarly high relationships were found for conscientiousness. These correlations for the subjects who rated themselves consistent are substantially higher than is normally found in personality research; whereas the correlations for the low consistency subjects are in the typically low range of +·2 to +·3. They also demonstrated that this approach can increase the predictive power of standard personality tests; they got the same subjects to complete the Eysenck Personality Inventory, and to rate themselves on how consistent they were on Extraversion. Again the correlations for the low variability subjects (average = ·51) were substantially higher than for the high variability subjects (average = ·22).

This emphasis on how the individual himself organizes his trait structure is fully compatible with an interactionist approach and is complementary to it. It is also consistent with a cognitive approach, in that traits may be adopted by an individual because of their value to him in helping to project a particular image of himself. He will concentrate only on those traits that are central to his self-concept and will, therefore,

121

monitor his behaviour in a restricted range of activities and settings.

To summarize: a behavioural or social learning viewpoint places a greater emphasis on the importance of learning and situations in personality. An interactionist view is taken, according to which human behaviour is seen as resulting from person variables, and situation variables, but especially the interaction between them. Previous learning, situational consistency and physical appearance are plausible alternatives to traits in accounting for behaviour consistencies. However, a revised conceptualization of traits in terms of idiosyncratic organization, incorporating the influences of previous learning, suggests a useful role for the concept.

Assessment for social learning

The social learning approach is not compatible with the traditional questionnaire method of assessing personality. The behaviourist's interest in the uniqueness of the individual disposes him more towards the assessment of behaviour where both person variables and situation variables can exert their full influence. Typically, behavioural assessments consist of an intensive assessment of a single individual in a restricted range of natural settings. Several different methods have been developed.

Direct observation
This refers to a simple count of the frequency of a particular kind of behaviour; for example, to assess friendliness we might count how often a person initiates conversation or smiles in a given setting. It is seldom possible to count the overall frequency of any particular behaviour; accordingly, a 'time-sampling' procedure is often used, in which the behaviours are recorded for say fifteen-minute periods randomly spaced throughout the day. With time-sampling, great care must be taken to cover times of the day which could affect the frequency of the behaviour of interest; for example, fre-

quency of social interaction may increase (or decrease) markedly at meal-times.

The importance of an extremely detailed or 'fine grain' analysis in the assessment of interpersonal behaviour has recently been demonstrated by Taylor (1975). In an observational study of nurse-patient interaction during a simple game of dominoes, she found that if one takes an overall look at the relationship between patient participation and nurse interaction, the more deteriorated patients received more prompts and an appropriate amount of verbal reinforcement. This is, of course, to be expected if nurse interventions are to have the effect of increasing the patient's effective participation. If, however, a more detailed examination is made of the temporal sequencing of prompts and reinforcements in relation to the patient's behaviour, a very different picture emerges. The nurse tends to prompt and reinforce irrespective of the patient's behaviour at that time. In other words, the nurse responds to her overall 'impression' of the patient and not to his actual behaviour. Only a minutely detailed analysis of the moment to moment interaction between patient and nurse could have elucidated this important information.

However, when we define precisely the nature of the behaviour being observed, we lose from our assessment those behaviours that may subsequently prove to be important, but which at the time fell outside our definition and were not recorded. Automated devices and coding systems are frequently used and enable a wider range of behaviours to be recorded.

Self-reports

A reliable way of assessment may simply be to ask a person to report on his own behaviour. A person may be asked to record how many cigarettes per day have been smoked, or how often he has experienced unpleasant thoughts; often the subject is asked to keep a diary of his behaviour. A major difficulty with self-recording is that often the simple act of recording itself has an effect on the behaviour; recording one's weight, for example, can result in a gradual weight loss.

123

Behavioural interviews

Interviewing techniques are ubiquitous and there is clearly nothing exclusively psychological or behavioural about them. Behavioural interviews differ, not in the techniques, but in the type of information explored. An outline of a behavioural interview for use in clinical situations has been given by Kanfer and Saslow (1969). Such items as the frequency, intensity, duration and situational determinants of problem behaviours are explored, together with such information as the effects of the problem behaviours on social, professional and domestic life, the reactions of significant people to the problem behaviours, the subject's skills and assets, his attempts at self-control and his major sources of reinforcement. The interview with the subject is often supplemented by interview with a relative or friend. The information obtained in the interviews is intended to lead directly to appropriate methods of treatment.

Survey schedules

Standardized inventories are often used to complement information obtained in interviews; they differ from conventional questionnaires in that they assess specific behaviours rather than enduring characteristics. A particular advantage of these schedules is that they sample a wide range of behaviours quickly and directly, and cover topics that might be inadvertently omitted, even by a skilled clinical interviewer. Cautela and Kastenbaum (1967), for example, have developed a list which can be used to identify the most appropriate reinforcers for use in clinical and other situations, using treatment techniques based on reinforcement principles. The subject is required to check the items (such as 'Tea', 'Folk Music', 'Newspapers', 'Clothes' or 'Making Love') by rating 'how much pleasure it gives you nowadays'.

Operant conditioning techniques

Operant conditioning techniques have been used directly in the assessment of human behaviour (for a review, see Weiss, 1968). For example, Nathan, Bull and Rossi (1968) have used

124

response frequency as a measure of attitude towards a therapist; the subjects had to press a lever at a given rate in order to maintain communication with the therapist via closed-circuit television. Peck (1973) used operant techniques to assess hearing range in a patient unable to communicate and for whom standard audiometric techniques were inappropriate. Such applications seem, however, to be of use in only a limited range of behaviours; the necessity for extensive instrumentation mostly precludes operant assessment in natural settings.

The use of behavioural assessment techniques does not of course imply that one can dispense with the normal methods of assessing measuring instruments; their reliability and validity must be examined in the same way as in other assessment procedure. In attempts to do this, problems have often been illuminated, bias on the part of the observer being a particularly difficult problem (Lipinski and Nelson, 1974). Nevertheless, it appears that a behavioural assessment generally leads to greater accuracy in predictions and may, therefore, be of greater utility in practical situations.

Applications of a social learning approach

There are few immediate practical applications of the social learning or behavioural approach to personality that can be said to derive specifically from it, as distinct from deriving from the behaviourist viewpoint in general. Many of the applications would still exist whether or not behaviourists had turned their attention to the domain of personality. However, the social learning approach to personality and behaviourism in general are so interwoven that it is appropriate to discuss applications in the context of this approach to personality.

A behaviourist would see no necessity to make out a special case for the onset and maintenance of different types of behaviour, whether 'normal' or 'abnormal'. It is of course acknowledged that all behaviours do not have their causes entirely in the environment (for example, the effects of brain

125

damage), but the same techniques may be used to change the behaviour, irrespective of its cause. They can be used in psychiatric hospitals or community houses to change 'abnormal' behaviour, and in classrooms or the home to develop 'normal' behaviour. The target behaviours and the appropriate reinforcers may differ, but the techniques are essentially the same. It is necessary, therefore, to examine the applications in just one area.

The main area of application is to emotional problems, and according to the precise techniques used, this application is known as behaviour therapy or behaviour modification. Only the former term will be used here, however, for the sake of simplicity. The various techniques used, and the assumptions underlying behaviour therapy are detailed in F3. It is, however, appropriate to introduce the general approach. Behaviour therapy is not a single technique but comprises a wide range of different techniques, the one used in any particular case being arranged only after a detailed analysis of the individual's problem. Two people may have problems that appear very similar, but they may be tackled by very different techniques because the environmental determinants of the problem may be different.

A major assumption is that learning processes play a vital role in the development and maintenance of problem behaviours. The application of learning principles should, therefore, be an effective way of alleviating problems. Some main groups of techniques will now be described. One group is used to help people with a phobia, or an incapacitating fear of an object or situation; the precise method will differ according to the particular problem, but typically it will consist of confronting the person with the feared object, in a graded approach when the person is relaxed, or it can be presented full-blown until the anxiety dissipates. The object may be presented in imagination, or in real-life if possible. Phobias of flying, of speaking in public or of leaving the house, for example, may be treated by these techniques.

A second group of techniques involves applying an aversive stimulus (such as electric shock) to someone when he indulges

126

in behaviour which he wishes to suppress; excessive drinking and gambling, and several forms of sexual problems such as exhibitionism have been tackled by these aversive methods. A third group uses the principles of operant conditioning, particularly reinforcement principles, and is applied especially to situations where it is desired to strengthen existing behaviours or to establish new behaviours. Helping severely handicapped patients to develop speech, or an obese person to stop over-eating are some applications of reinforcement principles. In addition, extensive use is made of modelling procedures and training people to apply reinforcement principles to themselves to develop self-control.

There are few areas of human problems that have not been alleviated by behaviour therapy techniques; other clinical problems include incontinence, self-injury, marital disharmony, tics, smoking and self-care in the handicapped. The same techniques have also been applied to educational problems (especially classroom behaviour) and even to medical conditions such as high blood pressure and migraine. Obviously behaviour therapy is not always the treatment of choice, but significant contributions have been made to the resolution of a wide range of problems, often in areas where there was no effective treatment available. (For a comprehensive review, see Agras, 1972.)

The investigation of outcome with any form of treatment is fraught with difficulties, as we saw with psychoanalytic treatments. Investigations of behaviour therapy are subject to similar difficulties, and again few well-controlled studies have been conducted and with many of them the subjects have been students with comparatively trivial problems, such as snake phobias, rather than hospital patients with incapacitating fears. For an approach that claims to have evolved from a rigorous based scientific basis, high quality reports are sparse, and behaviour therapists might usefully take more heed of the criticisms that many of their fellow behaviourists have raised against other treatment approaches. Nevertheless, the studies that have been done are almost uniformly supportive of behaviour therapy techniques. In particular, the effectiveness of

127

techniques based explicitly on operant conditioning principles is well-established, but even here it remains to be demonstrated that the changes are of long duration.

There are, therefore, still many unanswered questions in relation to the outcome of behaviour therapy. But in its short history of less than twenty years its proponents have demonstrated an encouraging degree of self criticism which has resulted in the development and refinement (and sometimes even the disuse) of several techniques. Much promise has been shown and there are signs that behaviour therapy will soon have much to offer in the alleviation of human suffering.

Evaluation

A behavioural approach to personality has been widely criticized. This approach could reasonably be considered as an alternative to personality theory, rather than a personality theory in itself; for many behaviourists, the term 'personality' refers to a field of study rather than an entity. A behavioural approach questions many of the traditional basic assumptions underlying personality research, and such revolutionary views will inevitably generate hostility. Skinner (1974) has discussed many of these issues, and defended the behavioural view of man.

Some of these criticisms are not directed towards behaviourism in general, but often towards the views of Skinner. A frequent criticism, for example, is that a behavioural approach does not attempt to provide an account of cognitive processes, but as we have seen, many theorists whose basic orientation is behavioural have examined cognitive processes in some detail and incorporate them in their theories (e.g. Mischel, 1973). Furthermore, Skinner himself acknowledges the importance of cognitive processes, but differs from others in his view of the appropriate ways of studying and interpreting them. Because of the present state of our knowledge, it may be useful and convenient to use such cognitive concepts as expectancies or 'meanings' of stimuli, but there should be nothing sacronsanct about them; they should be redefined or discarded where advances in knowledge render it necessary.

Although significant advances have been made, behaviourists consider that not enough is known to warrant the construction of a formal inclusive theory of human behaviour. Some psychologists prefer to investigate important and complex phenomena and construct comprehensive but largely conjectural theories about them. Behaviourists prefer (as do most scientists) to start with the simple and move gradually towards the complex. Accordingly, behavioural approaches are often accused of presenting an oversimplified view of human behaviour that cannot account for the complexities of everyday life. When human activities such as speech or creative writing (see A7) are formulated in behavioural terms, it appears superficial and unsatisfying and certainly does not do justice to the sum total of variables acting upon a person at any one time. Behaviourists have little in the way of an answer to such criticisms except to say that other approaches do not seem to provide a more satisfactory alternative account. Behavioural principles cannot explain all phenomena and behaviourists simply prefer to wait for (and work towards) the advent of technology that may eventually enable them to do so. In the meantime they can apply the knowledge that they have and feel reasonably confident about the security of its foundations.

There are, therefore, some cogent criticisms that can be levelled against a behavioural approach. Compared with other approaches, it does oversimplify and it does pay less attention to some important areas, but this is only because it is believed that the appropriate techniques are not yet available. A behavioural account of human behaviour is unsatisfactory, but behavioural psychology is sufficiently self-critical and has enough momentum to allow us to hope that this state of affairs may soon be rectified.

129

8
Conclusions

We are now in a position, having reviewed the major personality theories, to look back and see how the theories have fared in relation to the criteria for evaluating them that we discussed in the introductory chapter.

With the exception of the narrow-band and the interpersonal theories, all the theories have attempted to cover extensive areas of research and fields of application. However, there are inevitably weak points in each of the theories and certain areas are only dealt with sketchily. The type theory proposed by Eysenck is perhaps the most comprehensive in that it has been related to a vast range of phenomena, including socialization, conditioning and cortical activity.

Although they differ in emphasis all the theories are concerned with important areas of human activity such as the understanding and treatment of psychiatric disorders and educational problems. Apart from the psychoanalytic and field-dependency theories, they have all been sufficiently well organized and have stated their basic propositions with enough clarity to enable specific hypotheses to be derived and tested. Without exception they have all stimulated a great deal of research.

When we apply empirical criteria to the theories, the situation is much less satisfactory, in that most appear to be sadly

130

lacking in validity and utility. Few attempts are made to relate the theory to behaviour outside the psychological laboratory or clinic and when such attempts are made we find inconsistent results or at best only very weak relationships. Although evidence supporting the utility of theories in applied settings is generally hard to find, some theories have produced impressive results in limited areas. In particular, the applications of achievement motivation and the high reliability of field dependency are noteworthy. However, as we have seen, these results may often be explained equally well by other theories and the applications do not always appear to work for the reasons originally put forward. In clinical practice most personality measures have proved too unreliable to justify their use with individuals and attempts to assess personality change following psychotherapy have generally produced contradictory results.

The exception to these somewhat gloomy conclusions appears to be the behavioural-social-learning approach to personality. Findings in this area seem to be considerably more consistent and replicable. Also its value in applied settings is more firmly established than for any of the other theories. This probably reflects the fact that the more our predictions are based on data (such as questionnaires and self-report) which are far removed from actual behaviour in real-life settings, the less accurate our predictions are likely to be.

The current state of much personality trait research in relation to attitude change has been criticized by Elms (1972):

Research on the contribution of personality traits to attitude change is plentiful; it has, however, shown more promise than useful results. Perhaps this is because it seems so easy to do. Run a simple opinion change study; toss your volunteers whatever personality traits you have at hand, calculate a fast correlation – and presto! Hypomanic people are more persuasible than hypermanics. Or less. Or nonsignificantly different until you omit a few volunteers for very good reasons and run the correlation again by a different formula!

131

These comments could be applied equally forcefully to many other areas of personality research. The reasons for this unhappy state of affairs are probably highly complex. One suggestion put forward is that editors of psychological journals naturally have a preference for positive findings, or at least, for findings that are consistent with the hypothesis under investigation. Negative results probably have less chance of being submitted and accepted for publication. For this reason a small number of unrepresentative positive studies may be accorded undue significance by other researchers in the area. A second reason concerns the general reluctance of personality theorists to make changes in their theories in the light of contradictory evidence. Despite claims to scientific objectivity, theorists frequently become preoccupied with maintaining their own theories against criticism and rarely find it possible to take a fresh look at their own implicit assumptions. When negative findings are encountered the theorist can usually dismiss them by pointing to small faults in the design of the study, or by maintaining that the study did not involve a critical test of the theory. It is often possible to claim that the assessment devices used have been rendered obsolete by recent improvements, so that the criticisms do not apply to the new version of the test. Although it is desirable for the central elements in a theory to be resistant to frequent change, most theorists are extremely unwilling to admit a basic error and make extensive revisions of their theory. The result is that the field of personality has rarely seen the mutually advantageous interaction between research and theory construction so essential to scientific progress.

In Chapter 7 (social learning) we discussed a major area frequently neglected by personality theorists, viz. the effect of situations on behaviour. Although most theorists make passing reference to the importance of situational factors, they have invariably concentrated upon person variables in isolation. They have thus disregarded the evidence of sociologists, social psychologists and even the dictates of commonsense, all of which have repeatedly stressed the importance of taking into account the social context in any attempt to explain individual behaviour. These criticisms do not apply equally to all the theories discussed. Locus of Control, Need for Achievement and the Behavioural approaches in particular not only acknowledge the importance of environmental variables but all attempt to deal with them directly. It is significant that recent work of major importance in social psychology – Milgram's studies on *Obedience to Authority* – was conducted without any recourse to the traditional concepts of personality and efforts to utilize personality trait measures in the prediction of 'obedience' have so far been unsuccessful.

Mischel (1973) has argued forcefully that an adequate account of personality must embrace three perspectives, viz. situational determinants, person variables and experiential phenomena (that is, the individual's subjective interpretation of events). None of the theories discussed is satisfactory on all three counts. However, it is suggested that the behavioural approach, buttressed by some of the central concepts of the interpersonal theorists (particularly Kelly) offers the most satisfactory way forward. In addition, biological factors of the type favoured by Eysenck may ultimately have to be given greater prominence.

There is little doubt that much could still be achieved by personality researchers continuing to work within the confines of traditional theories. The purpose of the synthesis proposed here is not to achieve some kind of spurious eclecticism but to recognize the complex realities of the study of personality. In this way personality research may at long-last fulfil its promise to make a significant contribution to important human problems.

Further Reading

Bannister, D. and Mair, M. J. M. (1968) *The Evaluation of Personal Constructs*. London: Academic Press Inc.

Bergin, A. E. and Garfield, S. L. (1971) *Handbook of Psychotherapy and Behaviour Change: an empirical analysis*. New York: Wiley.

Brody, N. (1972) *Personality Research and Theory*. London: Academic Press Inc.

Cattell, R. B. (1965) *The Scientific Analysis of Personality*. Harmondsworth: Penguin.

Child, D. (1973) *The Essentials of Factor Analysis*. London: Holt, Rinehart and Winston.

Eysenck, H. J. and Eysenck, S. B. G. (1969) *The Structure and Measurement of Personality*. London: Routledge and Kegan Paul.

Jones, E. (1964) *The Life and Work of Sigmund Freud*. Harmondsworth: Penguin.

Krasner, L. and Ullmann, L. P. (1973) *Behaviour Influence and Personality: the social matrix of human action*. New York: Holt, Rinehart and Winston.

Levy, L. M. (1970) *Conceptions of Personality: Theory and Research*. New York: Random House.

Pervin, L. A. (1970) *Personality: Theory, Assessment, and Research*. New York: Wiley.

Rogers, C. R. (1959) A theory of therapy, personality, and interpersonal relationships as developed in the client-centred framework. In S. Koch (ed.) *Psychology: A Study of a Science*, Vol. 3. New York: McGraw-Hill, 184–256.

Rotter, J. B., Chance, J. E. and Phares, E. J. (1972) *Applications of a Social Learning Theory of Personality*. New York: Holt, Rinehart and Winston.

References
and Name Index

The numbers in italics following each entry refer to page numbers within this book.

Abramowitz, S. I. (1969) Locus of control and self-reported depression among college students. *Psychological Reports* 25: 149–50. *81*

Agras, S. (1972) *Behavior Modification: Principles and Clinical Applications.* Boston: Little Brown and Co. *127*

Allen, B. P. and Potkay, C. R. (1973) Variability of self-description on a day-to-day basis: longitudinal use of the adjective generation technique. *Journal of Personality* 41: 638–52. *112*

Allport, G. (1937) *Personality: A Psychological Interpretation.* New York: Holt. *120*

Allport, G. W. (1961) *Pattern and Growth in Personality.* New York: Holt, Rinehart and Winston. *10*

Asch, S. E. (1958) Effects of group pressure upon the modification and distortion of judgements. In E. Maccoby *et al.* (eds) *Readings in Social Psychology.* New York: Holt, Rinehart and Winston. *15*

Atkinson, J. W. and Feather, N. T. (1966) *A Theory of Achievement Motivation.* New York: Wiley. *97*

Bandura, A. (1969) *Principles of Behavior Modification.* London: Holt, Rinehart and Winston. *117*

Bannister, D. (1966) Psychology as an exercise in paradox. *Bulletin of the British Psychological Society* 19: 21–6. *37*

Bannister D. and Fransella F. (1967) *Grid Test of Schizophrenic Thought Disorder Manual.* London: Barnstaple Psychology Test Publications. *51*

Bannister, D., Fransella, F. and Agnew J. (1971) Characteristics and validity of the Grid Test of Thought Disorder. *British Journal of Social and Clinical Psychology* 10: 144–51. *52*

Bannister, D. and Mair, J. M. M. (1968) *The Evaluation of Personal*

135

Constructs. London: Academic Press. *49, 51*

Bem, D. J. and Allen, A. (1974) On predicting some of the people some of the time: a search for cross situational consistencies in behavior. *Psychological Review 81*: 506–20. *120*

Blackburn, R. (1969) Sensation seeking, impulsivity, and psychopathic personality. *Journal of Consulting and Clinical Psychology 33*: 571–4. *107*

Blackman, D. (1974) *Operant Conditioning: An Experimental Analysis of Behaviour*. London: Methuen. *115*

Bonarius, J. C. J. (1970) Fixed role therapy: a double paradox. *British Journal of Medical Psychology 43*: 213–19. *52*

Bowers, K. S. (1973) Situationism in psychology: an analysis and a critique. *Psychological Review 80*: 307–36. *118, 119*

Brody, N. (1972) *Personality Research and Theory*. London: Academic Press Inc. *35, 67, 73*

Bruner, J. S. (1956) You are your constructs. *Contemporary Psychology 1*: 355–6. *54*

Butler, R. A. (1958) Exploratory and related behaviour: A new trend in animal research. *Journal of Individual Psychology 14*: 111–20. *104*

Candy, J., Balfour, F. M. G., Cawley, R. H., Hildebrand, H. P., Malan, D. H., Marks, I. M. and Wilson, J. (1972) A feasibility study for a controlled trial of formal psychotherapy. *Psychological Medicine 2*: 345–62. *32*

Cattell, R. B. (1963) Theory of fluid and crystallised intelligence: a critical experiment. *Journal of Educational Psychology 54*: 1–22. *66*

Cattell, R. B. (1965) *The Scientific Analysis of Personality*. Harmondsworth: Penguin. *61, 62, 65*

Cattell, R. B. (1974) How good is the modern questionnaire? General principles for evaluation. *Journal of Personality Assessment 38*: 115–29. *64*

Cattell, R. B., Eber, H. W. and Tatsuoka, M., (1970) *Handbook for the Sixteen Personality Factor Questionnaire*. Champaign, Ill.: Institute for Personality and Ability Testing. *62, 63*

Cautela, J. R. and Kastenbaum, R. (1967) A reinforcement survey schedule for use in therapy, training and research. *Psychological Reports 20*: 1115–30. *124*

Claridge, G. S. (1967) *Personality and Arousal*. Oxford: Pergamon Press. *72*

Claridge, G. S. and Chappa, H. J. (1973) Psychoticism: a study of its biological basis in normal subjects. *British Journal of Social and Clinical Psychology 12*: 175–87. *75*

Cochrane, R. (1974) Crime and personality: theory and evidence. *Bulletin of the British Psychological Society 27*: 19–22. *77*

Collins, B. E. (1974) Four components of the Rotter internal-external scale: belief in a difficult world, a just world, a predictable world, and a politically repressive world. *Journal of Personality and Social Psychology 29*: 381–91. *83*

Elms, A. C. (1972) *Social Psychology and Social Relevance*. Boston: Little Brown and Co. *131*

Epstein, S. (1973) The self-concept revisited: or a theory of a

theory. *American Psychologist 28*: 404–16. *45*
Eysenck, H. J. (1947) *Dimensions of Personality*. London: Routledge and Kegan Paul. *69*
Eysenck, H. J. (1952) *The Scientific Study of Personality*. London Routledge and Kegan Paul. *70*
Eysenck, H. J. (1953) *The Structure of Human Personality*. London: Methuen. *68*
Eysenck, H. J. (1964) *Crime and Personality*. London: Routledge and Kegan Paul. *76*
Eysenck, H. J. (1965) *Fact and Fiction in Psychology*. Harmondsworth: Penguin. *69, 71*
Eysenck, H. J. (1967) *The Biological Basis of Personality*. Springfield, Ill.: C. C. Thomas. *71*
Eysenck, H. J. (1970) A dimensional system of psychodiagnostics. In A. Mahrer (ed.) *New approaches to Personality Classification.* New York: Columbia University Press. *75, 76*
Eysenck, H. J. (1972) The experimental study of Freudian concepts. *Bulletin of the British Psychological Society 25*: 261–7. *34*
Eysenck, H. J. (1974) Crime and personality reconsidered. *Bulletin of the British Psychological Society 27*: 23–4. *77*
Eysenck, H. J. and Eysenck, S. B. G. (1964) *Manual of the Eysenck Personality Inventory*. London: University of London Press. *73*
Eysenck, H. J. and Eysenck, S. B. G. (1969) *The Structure and Measurement of Personality*. London: Routledge and Kegan Paul. *64*
Eysenck, S. B. G. and Eysenck, H. J. (1963) The validity of questionnaires and rating assessment of extraversion and neuroticism and their factorial stability. *British Journal of Psychology 54*: 51–62. *74*
Eysenck, S. B. G. and Eysenck, H. J. (1970) Crime and personality: an empirical study of the three-factor theory. *British Journal of Criminology 10*: 225–39. *77*
Eysenck, S. B. G., Eysenck, H. J. and Shaw, L. (1974) The modification of personality and lie scale scores by special 'honesty' instructions. *British Journal of Social and Clinical Psychology 13*: 41–50. *74*
Festinger, J., Schachter, S. and Back, K. (1950) *Social Pressures in Informal Groups*. New York: Harper. *17*
Fiske, D. W. and Goodman, G. (1965) The post therapy period. *Journal of Abnormal Psychology 70*: 169–79. *44*
Foulds, G. A. (1973) Has anybody here seen Kelly? *British Journal of Medical Psychology 46*: 221–5. *55*
Freud, A. (1936) *The ego and the mechanisms of defense*. London: Hogarth. *24*
Gardner, R. W., Holzman, P. S., Klein, G. S., Linton, H. B. and Spence, D. P. (1959) Cognitive control: a study of individual consistencies in cognitive behavior. *Psychological Issues 1*: No. 4. *87*
Garfield, S. L., Praner, R. A. and Bergin, A. E. (1971) Evaluation of outcome in psychotherapy. *Journal of Consulting and Clinical Psychology 37*: 307–13. *42, 44*
Gathercole, C. E., Bromley, E. and Ashcroft, J. B. (1970) The

reliability of repertory grids. *Journal of Clinical Psychology* 26: 513–16. *51*

Gibson, H. B. (1971) The validity of the Eysenck Personality Inventory studied by a technique of peer-rating item by item, and by socio-metric comparisons. *British Journal of Social and Clinical Psychology* 10: 213–20. *74*

Goldberg, L. R. (1971) A historical survey of personality scales and inventories. In P. McReynolds (ed.) *Advances in Psychological Assessment* 2. Palo Alto, Calif. Science and Behavior Books. *20*

Goldstein, G., Neuringer, C., Reiff, C. and Shelly, C. H. (1968) Generalizability of field dependency in alcoholics. *Journal of Consulting and Clinical Psychology* 32: 560–4. *91*

Guntrip, H. J. S. (1971) *Psychoanalytic Theory, Therapy, and the Self*. London: Hogarth Press. *27*

Gurin, P., Gurin, G., Lao, R. and Beattie, M. (1969) Internal–external control in the motivational dynamics of Negro youth. *Journal of Social Issues* 25: 29–53. *83*

Hall, C. S. and Lindzey, G. (1957) *Theories of Personality*. New York: Wiley. *14, 18, 56*

Harris, S. L. and Nathan, P. E. (1973) Parents' locus of control and perception of cause of children's problems. *Journal of Clinical Psychology* 29: 182–4. *82*

Harrow, M. and Ferrante, A. (1969) Locus of control in psychiatric patients. *Journal of Consulting and Clinical Psychology* 33: 582–9. *84*

Hartshorne, H. and May, M. A. (1928) *Studies in the Nature of Character. Volume I. Studies in Deceit*. New York: Macmillan. *110*

Heim, A. (1970) *Intelligence and Personality: Their Assessment and Relationship*. Harmondsworth: Penguin. *73*

Holmes, D. S. (1974) The conscious control of thematic projection. *Journal of Consulting and Clinical Psychology* 42: 323–9. *29*

Holzman, P. S. (1974) Personality. *Annual Review of Psychology* 25: 247–77. *110*

Howarth E. and Browne, A. (1971) An item-factor analysis of the 16 P.F. *Personality* 2: 117–39. *64*

Hull, C. L. (1943) *Principles of Behavior*. New York: Appleton-Century-Crofts. *71*

Hurley, J. R. (1971) Science and fiction in executive training. *Journal of Applied Behavioral Science* 7: 230–3. *102*

Joe, V. C. (1971) Review of the internal–external control construct as a personality variable. *Psychological Reports* 28: 619–40. *83*

Jones, E. E. and Nisbett, R. E. (1971) *The Actor and Observer: Divergent Perceptions of the Causes of Behavior*. New York: General Learning Press. *113*

Kanfer, F. H. and Saslow, G. (1969) Behavioral diagnosis. In C. M. Franks (ed.) *Behavior Therapy: Appraisal and Status*. New York: McGraw-Hill. *124*

Kelley, H. H. (1950) The warm–cold variable in first impressions of persons. *Journal of Personality* 18: 431–9. *16*

Kelly, G. A. (1955) *The Psychology of Personal Constructs*. New York: Norton. *47*

Kernberg, O. F. (1974) Summary and conclusions of psychotherapy and phychoanalysis. Final report of the Menninger Foundation's Psychotherapy Research Project. In H. H. Strupp et al. (eds) Psychotherapy and Behavior Change 1973. Chicago: Aldine. 32

Kish, G. B. (1966) Studies of sensory reinforcement. In W. K. Honig (ed.) Operant Behavior: Areas of Research and Application. New York: Appleton-Century-Crofts. 104

Kish, G. B. (1970) Reduced cognitive innovation and stimulus seeking in chronic schizophrenia. Journal of Clinical Psychology 26: 170-4. 107

Kline, P. (1972) Fact and Fantasy in Freudian Theory. London: Methuen. 34

Kolb, D. A. (1965) Achievement motivation training for underachieving high school boys. Journal of Personality and Social Psychology 2: 763-92. 101, 103

Konstadt, N. and Forman, E. (1965) Field dependence and external directedness. Journal of Personality and Social Psychology 1: 490-3. 88

Lambley, P. and Silbowitz, M. (1973) Rotter's internal–external scale and prediction of suicide contemplators among students. Psychological Reports 33: 585-6. 85

Levy, L. H. (1970) Conceptions of Personality. Theory and Research. New York: Random House. 12

Linton, H. B. (1955) Dependence on external influence: correlates in perception, attitudes and judgement. Journal of Abnormal and Social Psychology 51: 502-7. 88

Lipinski, D. and Nelson, R. (1974) Problems in the use of naturalistic observation as a means of behavioral assessment. Behavior Therapy 5: 341-51. 125

Looft, W. R. and Baranowski, M. D. (1971) An analysis of five measures of sensation seeking and preference for complexity. Journal of General Psychology 85: 307-13. 107

Luchins, A. S. and Luchins, E. H. (1967) Conformity: task vs. social requirements. Journal of Social Psychology 71: 95-105. 15

Lykken, D. G. (1971) Multiple factor analysis of personality research. Journal of Experimental Research in Personality 5: 161-70. 66, 67

McAllister, L. W. (1970) Modification of performance on the Rod-and Frame Test through token reinforcement procedures. Journal of Abnormal Psychology 75: 124-30. 94

McClelland, D. C. (1961) The Achieving Society. Princeton: Van Nostrand. 98

McClelland, D. C. (1972) What is the effect of Achievement Motivation training in the Schools? Teachers College Record 74: 129-45. 101

McClelland, D. C., Atkinson, J. W., Clark, R. A. and Lowell, E. L., (1953) The Achievement Motive. New York: Appleton-Century-Crofts. 98, 99

McClelland, D. C. and Winter, D. G. (1969) Motivating Economic Achievement. New York: Free Press. 101

McGeorge, C. (1974) Situational variation in level of moral judgement. British Journal of Educational Psychology 44: 116-22. 111

McGuire, R. J., Mowbray, R. M. and Vallance, R. C. (1963) The

Maudsley personality inventory used with psychiatric patients. *British Journal of Psychology 54*: 157–66. *76*

Mehrabian, A. and Russell, J. A. (1973) A measure of arousal seeking tendency. *Environment and Behavior 5*: 315–33. *107*

Milgram, S. (1974) *Obedience to Authority: an experimental overview.* London: Tavistock. *133*

Minard, J. G. and Mooney, W. (1969) Psychological differentiation and perceptual defense and studies of the separation of perception from emotion. *Journal of Abnormal and Social Psychology 74*: 131–9. *93*

Mischel, T. (1964) Personal constructs, rules and the logic of clinical activity. *Psychological Review 71*: 180–92. *54*

Mischel, W. (1968) *Personality and Assessment.* New York: Wiley. *110, 111*

Mischel, W. (1972) Direct versus indirect personality assessment: evidence and implications. *Journal of Consulting and Clinical Psychology 38*: 319–24. *111*

Mischel, W. (1973) Toward a cognitive social learning reconceptualization of personality. *Psychological Review 80*: 252–83. *118, 119, 128, 133*

Morris, R. J. (1973) Factors affecting occurrence of experimental repression. *Psychological Reports 33*: 147–56. *34*

Nathan, P. E., Bull, T. A. and Rossi, A. M. (1968) Operant range and variability during psychotherapy: description of possible communication signatures. *Journal of Nervous and Mental Diseases 146*: 41–9. *124*

Pavlov, I. P. (1927) *Conditioned Reflexes.* London: Oxford University Press. *71*

Peck, D. F. (1973) Operant assessment of hearing range in a nonverbal retardate. *Behavior Therapy 4*: 319–20. *125*

Pervin, L. A. (1970) *Personality: Theory, Assessment and Research.* New York: Wiley. *10*

Poole, A. D. (1973) Some psychiatric correlates of personal construct systems. Unpublished Ph.D. Thesis: University of Dublin. *52*

Rachman, S. (1971) *The Effects of Psychotherapy.* Oxford: Pergamon. *32*

Radley, A. R. (1974) Schizophrenic thought disorder and the nature of personal constructs. *British Journal of Social and Clinical Psychology 13*: 315–27. *52*

Rogers, C. R. (1940) The process of psychotherapy. *Journal of Consulting Psychology 4*: 161–4. *43*

Rogers, C. R. (1951) *Client-centred Therapy.* Boston Mass.: Houghton. *40*

Rogers, C. R. (1957) The necessary and sufficient conditions of therapeutic personality change. *Journal of Consulting Psychology 21*: 87–9. *43*

Rogers, C. R. (1959) A theory of therapy, personality, and interpersonal relationships as developed in the client-centred framework. In S. Koch (ed.) *Psychology: A Study of a Science.* Vol. 3. New York: McGraw-Hill: 184–256. *40*

Rotter, J. B. (1954) *Social Learning and Clinical Psychology.* Englewood Cliffs: Prentice-Hall. *79*

Rotter, J. B. (1966) Generalised expectancies for internal versus external control of reinforcement. *Psychological Monographs 80*: No. 1 (Whole No. 609). *82, 84*

Rump, E. E. and Court, J. (1971) The Eysenck Personality Inventory and social desirability response set with student and clinical groups. *British Journal of Social and Clinical Psychology 10*: 42–54. *74*

Schneider, D. J. (1973) Implicit personality theory: a review. *Psychological Bulletin 79*: 294–309. *112*

Sears, R. R. (1944) Experimental analyses of psychoanalytic phenomena. In J. McV. Hunt (ed.) *Personality and the Behavior Disorders 1*. New York: Ronald Press. *33*

Seligman, M. (1973) Fall into helplessness. *Psychology Today 7*: 43–8. *81*

Skinner, B. F. (1938) *The Behavior of Organisms*. New York: Appleton-Century-Crofts. *116*

Skinner, B. F. (1974) *About Behaviorism*. London: Jonathan Cape. *114, 128*

Stengel, E. (1963) Hughlings Jackson's influence on psychiatry. *British Journal of Psychiatry 109*: 348–55. *35*

Stephenson, W. (1953) *The Study of Behavior*. Chicago, Ill.: University of Chicago Press. *41*

Taylor, V. A. (1975) A behavioural analysis of activity group therapy with hard-core schizophrenics. Unpublished Doctoral Thesis: University of Aberdeen. *123*

Truax, C. B. (1966) Reinforcement and non-reinforcement in Rogerian psychotherapy. *Journal of Abnormal and Social Psychology 71*: 1–9. *45*

Truax, C. B. and Carkhuff, R. R. (1967) *Toward Effective Counseling and Psychotherapy*. Chicago: Aldine. *44*

Truax, C. B., Schuldt, W. J. and Wargo, D. G. (1968) Self-ideal concept congruence and improvement in group psychotherapy. *Journal of Consulting and Clinical Psychology 32*: 47–53. *42*

Ulrich, R. E., Stachnik, T. J. and Stainton, N. R. (1963) Student acceptance of generalized personality interpretations. *Psychological Reports 13*: 831–4. *17*

Vernon, P. E. (1964) *Personality Assessment: A critical survey*. New York: Wiley. *29*

Vernon, P. E. (1972) The distinctiveness of field independence. *Journal of Personality 40*: 366–91. *89, 90, 93*

Wachtel, P. L. (1972) Field dependence and psychological differentiation: re-examination. *Perceptual and Motor Skills 35*: 179–89. *92, 93, 94*

Warr, P. B. (1964) Proximity as a determinant of positive and negative sociometric choice. *British Journal of Social and Clinical Psychology 4*: 104–9. *17*

Waters, L. W., and Waters, L. K. (1969) Relationships between a measure of 'sensation-seeking' and Personal Preference Schedule need scales. *Educational and Psychological Measurement 29*: 983–5. *109*

Weinstein, M. S. (1969) Achievement motivation and risk preference.

Journal of Personality and Social Psychology 13: 153-72. *100, 120*

Weiss, R. L. (1968) Operant conditioning techniques in psychological assessment. In P. McReynolds (ed.) *Advances in Psychological Assessment.* 1 Palo Alto, California. Science and Behavior Books. *124*

Whyte, L. L. (1962) *The Unconscious before Freud.* London: Tavistock. *35*

Williams, C. B. and Nickels, J. B. (1969) Internal-external control dimension as related to accident and suicide proneness. *Journal of Consulting and Clinical Psychology 33*: 485-94. *85*

Williams, J. D., Dudley, H. K. and Overall, J. E. (1972) Validity of the 16 P.F. and the M.M.P.I. in a mental hospital setting. *Journal of Abnormal Psychology 80*: 261-70. *65*

Witkin, H. A. (1965) Psychological differentiation and forms of pathology. *Journal of Abnormal and Social Psychology 70*: 317-36. *90, 92*

Witkin, H. A., Goodenough. D. R. and Karp, S. A. (1967) Stability of cognitive style from childhood to young adulthood. *Journal of Personality and Social Psychology 7*: 291-300. *89, 95*

Witkin, H. A., Lewis, H. B. and Weil, E. (1968) Affective reactions and patient-therapist interactions among more differentiated and less differentiated patients early in therapy. *Journal of Nervous and Mental Diseases 146*: 193-208. *91*

Wright, D. S., Taylor, A., Davies, D. R., Sluckin, W., Lee, S. G. M. and Reason, J. T. (1970) *Introducing Psychology. An Experimental Approach.* Harmondsworth: Penguin. *10*

Zubek, J. P. (1969) *Sensory Deprivation: Fifteen Years of Research.* New York: Appleton-Century-Crofts. *105*

Zuckerman, M. (1971) Dimensions of sensation seeking. *Journal of Consulting and Clinical Psychology 36*: 45-52. *106*

Zuckerman, M., Bone, R. N., Neary, R., Mangelsdorff, D. and Brustman, B. (1972) What is the sensation-seeker? Personality trait and experience correlates of the sensation-seeking scales. *Journal of Consulting and Clinical Psychology 39*: 308-21. *108*

Zuckerman, M., Kolin, E. A., Price, L. and Zoob, I. (1964) Development of a sensation-seeking scale. *Journal of Consulting and Clinical Psychology 28*: 477-82. *106*

Zuckerman, M. and Link, K. (1968) Construct validity for the sensation-seeking scale. *Journal of Consulting and Clinical Psychology 32*: 420-6. *109*

Zuckerman, M., Schultz, D. P. and Hopkins, T. R. (1967) Sensation seeking and volunteering for sensory deprivation and hypnosis experiments. *Journal of Consulting Psychology 31*: 358-63. *107*

Subject Index

143

For Product Safety Concerns and Information please contact our EU
representative GPSR@taylorandfrancis.com Taylor & Francis Verlag GmbH,
Kaufingerstraße 24, 80331 München, Germany

Printed and bound by CPI Group (UK) Ltd, Croydon, CR0 4YY
01/05/2025
01858518-0001